EFFECTIVENESS OF BANK FILTRATION FOR WATER SUPPLY IN ARID CLIMATES: A CASE STUDY IN EGYPT

T0136176

AHMED RAGAB ABDELRADY MAHMOUD

EFFECTIVENESS OF BANK FILTRATION FOR WATER SUPPLY IN ARID CLIMATES: A CASE STUDY IN EGYPT

DISSERTATION

Submitted in fulfilment of the requirements of
the Board for Doctorates of Delft University of Technology
and
of the Academic Board of the IHE Delft
Institute for Water Education
for
the Degree of DOCTOR
to be defended in public on
Tuesday 10 November, 2020, at 10:00 hours
in Delft, the Netherlands

by

Ahmed Ragab Abdelrady MAHMOUD

Master of Science in Water Resources and Environment management

International Institute for Geo-Information Sciences and Earth Observation (ITC),

University of Twente, the Netherlands

born in Aswan, Egypt

This dissertation has been approved by the

Promotor: Prof.dr. M.D. Kennedy and
copromotor: Dr.ir. S.K. Sharma

Rector Magnificus TU Delft	Chairman
Rector IHE Delft	Vice-Chairman
Prof.dr. M.D. Kennedy	IHE Delft / TU Delft, promotor
Dr.ir. S.K. Sharma	IHE Delft, copromotor

Independent members:

Prof.dr. P.J. Stuijfzand	TU Delft
Prof.dr.ir. L.C. Rietveld	TU Delft
Prof.dr.-ing. T. Grischeck	Hochschule für Technik und Wirtschaft Dresden, Germany
Prof.dr. A.K. Moawad	Al-Azhar University, Egypt
Prof.dr. M.E. McClain	IHE Delft / TU Delft, reserve member

This research was conducted under the auspices of the Graduate School for Socio-Economic and Natural Sciences of the Environment (SENSE)

CRC Press/Balkema is an imprint of the Taylor & Francis Group, an informa business
© 2020, Ahmed Mahmoud

Published by:
CRC Press/Balkema
Schipholweg 107C, 2316 XC, Leiden, the Netherlands
Pub.NL@taylorandfrancis.com
www.crcpress.com – www.taylorandfrancis.com

ISBN 978-0-367-74673-5

This thesis is dedicated to my family ...

ACKNOWLEDGMENTS

The primary purpose of this thesis was to contribute to providing drinking water of adequate quality for the people in my country (Egypt) and worldwide who suffer from a shortage of safe water. I hope that this work could help in achieving this goal. This work was funded by The Netherlands Fellowship Programme NFP and I am deeply thankful for this support.

Many people have assisted and motivated me to do this work in the last few years. I would like to express my appreciation and gratitude to them.

I would like to express my sincere appreciation to Prof. Maria Kennedy, my promoter, who provided me with the opportunity to pursue my PhD at IHE-Delft. Thank you for your guidance, advice and encouragement in finalizing my research and keeping me on the right track. I owe Dr. Saroj Sharma, my sincere thanks for his insightful contribution, effective comments and suggestions during the study. Without your conscientious support and enlightening feedback, the research would not have been carried out. I am deeply grateful to my supervisor Dr. Ahmed Sefelnasr, who has guided me in the development of a groundwater model for Aswan City, for his precious advice, and guidance and valuable discussions which has a vast contribution to this thesis.

I heartily appreciate and acknowledge the support of UNESCO-IHE Environmental Engineering Laboratory staff (Fred Kruis, Frank Wiegman, Berend Lolkema, Ferdi Battes, Peter Heerings, and Lyzette Robbemont) in conducting the laboratory experiments and analysis. I am deeply grateful for Jolanda Boots (Ph.D. Fellowship and Admission Officer), Ellen de Kok, Anique Karsten and Floor Felix for their support during my study. I am also thankful to Prof. Dr. Piet Lens for his valuable comment during my PhD proposal defence. My gratitude also goes to Dr. Nirajan Dhakal, Dr. Paolo Paron, Dr. Tibor Stigter, and Dr. Claire Taylor for their support. It was my pleasure to supervise three master students (Abdullahi Ahmed, Jovine Bachwenkizi and Abubakar Ubale), thanks for your contributions.

During this study, field and laboratory studies were conducted at Egyptian Water and Wastewater Companies; I would like to seize this opportunity to express my deep gratitude and appreciation to my dear colleagues at Aswan Company (Atef Abdelbaset, Ahmed Abdo, Hamza Ahmed, Mohamed Saad, Ahmed Mahmoud, and Mohamed Ali) Environmental affairs agency (Ahmed Negm and Mahmoud Hassan), Holding Company (Dr. Mohamed Bakr- manager of the reference laboratory- Dr. Mohamed Sayed, Amr Abogabal, Ahmed Saad, Mohamed Hossam A. Ghaly, Dr. Mohamed Barakat, Mahmoud Gomaa and Mohamed Mossad), Assiut University (Ahmed Abdelmagssoad, Momen Mohamed and Adnan Osama), Minia University (Dr. Mustafa Elrawy), and Aswan

University (Dr. Ali Hemdan, Dr. Omr Hamdi and Dr. Adel Abdelfahem). Special thanks go to Prof. Sayed Abdo, who passed away in 2017, for his support and advice, may Allah forgive him.

Gratitude also goes to my colleagues at IHE-Delft who facilitated my social life during my thesis. Many thanks to Mohanad Abunada, Taha Al-Washali, Almotasembellah Abushaban, Ahmed Elghandour, Mohaned Sousi, Mary Barrios Hernandez, Marmar Ahmed, Hisham Elmilady, Muhammad Nasir, Musaed Aklan, Shakeel Hayat, Shaimaa Theol, Aftab Nazeer, Jakia Akter, Reem Digna, Khalid Hassaballah, Shahnoor Hasan, Jessica Salcedo and Zahrah Musa. I am thankful to my Egyptian colleagues (Basem Mahmoud, Moahmed Essam, Mohamed Gohenim, Shady Moahmed, Mohamed Hamed, Mahmoud Abdlbaky, Abdelrahamn Mohamed) and special thank go to Ali Obied for his contribution to this work.

Last March, I started my postdoc research and became a member of a team that investigates the transport of environmental DNA-particles in natural water systems. I would like to take this opportunity to thank Dr. Thom Bogaard, Dr. Jan Willem, Ali Ben Hadi, Bahareh Kianfar, Swagatam Chakraborty, Coco Tang, Sulalit Bandyopadhyay and Zina Al-Saffar. This period was enjoyable and full of insights.

Above all, I am extremely grateful to my parents and my family members for their support, encouragement, love, prayers, advice and sacrifices, I dedicate this thesis as an insignificant gift for your endless support, and I owe you my life. Many thanks to my brother (Ayman) and sisters (Eman and Shimaa), their children (Roaa, Ziad and Ans). I am deeply grateful to my wife (Ola), her family (Dr. Barakat, Ahmed, Mohamed, Yara, Basmala and their mother) and my children (Omr and Youssef), thanks for your support. To my wife, thank you for taking care of our children in the last few years, you are everything to me and I am so fortunate to have you by my side. Finally, many thanks to everyone who contributed to this work.

Ahmed Mahmoud

November, 2020

Delft, Netherlands

SUMMARY

Pollution of surface water resources is a growing worldwide problem and conventional water treatment technologies generally cannot effectively remove the wide variety of contaminants from surface water sources. Natural treatment techniques such as bank filtration (BF) may represent a cost-effective and sustainable alternative compared to advanced treatment technologies. BF is achieved by the continuous pumping of a hydraulically linked well, situated in the vicinity of a surface water supply, inducing filtration flow to the interception aquifer. The subsurface sediments serve as natural barriers capable of eliminating various types of pollutants from water bodies. The efficacy of this technique has been demonstrated in developed countries for more than a hundred years as a means of producing high-quality drinking water. However, the environmental conditions in certain countries are expected to vary as a result of climate change in the coming years. Equally, several arid-zone countries (e.g., Egypt and India) have recently paid more attention to BF as a means of meeting their demand for drinking water. The efficiency of BF in arid conditions still needs to be investigated. These regions are characterised by (i) highly variable hydrological parameters and poor surface water infiltration into the adjacent aquifers. As a consequence, long travel times may be required and the potential for anaerobic conditions to form increases; (ii) high temperatures, which promote biological activity along the infiltration path affecting BF treatment mechanisms; and (iii) surface waters that are highly polluted, particularly organic substances that significantly influence the reactions taking place in the infiltration zone. BF is a site-specific technique, and there is a lack of knowledge on its design and sustainable management in arid and polluted environments. This thesis aimed to analyse the efficacy of BF in eliminating chemical contaminants and supplying good quality drinking water to meet local environmental standards.

Dissolved organic matter (DOM) is very important in terms of the physical processes and biochemical reactions that occur in the infiltration zone, and is a significant determinant of bank filtrate quality. This research aimed to assess the behaviour of organic matter fractions during the filtration process using analytical techniques such as size exclusion liquid chromatography with organic carbon detection and organic nitrogen detection (LC-OCD/OND), and fluorescence excitation-emission matrix spectroscopy integrated with parallel factor analysis (FEEM-PARAFAC). Laboratory-scale batch and soil column experiments were performed to examine the impact of environmental parameters (i.e., temperatures between 20–30 °C and redox conditions) on the attenuation of organic constituents in the filtration region. Experiments were conducted using different types of feed water collected from different sources. Results showed that non-humic compounds (e.g., biopolymers) were most prone to elimination during infiltration. The removal of biopolymers exceeded 80% under aerobic conditions regardless of the ambient temperature. Conversely, humic compounds displayed temperature-dependent behaviour with preferential removal at lower temperatures (20–25 °C). However, an increase in humic compounds in the effluent was observed at high temperatures (30 °C), which might be attributed to (i) the dissolution of soil organic matter into the infiltrated waterand (ii) the presence of micro-organisms capable of turning labile compounds into more refractory ones. Organic fluorescence analysis revealed an increase in the concentrations of humic compounds of terrestrial origin in the effluent at higher temperatures.

BF has been shown to have a relatively high potential to remove organic micropollutants (OMPs) in developed countries. However, its effectiveness under arid conditions is still not thoroughly understood. To address this gap, a laboratory-scale batch study was conducted to investigate the behaviour of different OMPs in the filtration region. The removal efficiencies of various classes of OMPs (polyaromatic hydrocarbons, insecticides, and pesticides) under varying conditions of raw water sources, temperature (20–30 °C), and redox conditions (i.e., oxic, anoxic, and anaerobic) were examined for a residence time of 30 days. Results revealed that the attenuation of OMPs and their reactions with the surrounding environment is strongly controlled by their properties (i.e., solubility). Highly-hydrophobic OMPs (i.e., DDT, pyriproxyfen, pendimethalin, β-BHC, endosulfan sulphate, and PAHs; logS < -4) tended to adsorb (> 80%) onto sand grains regardless of the temperature and redox conditions. Hydrophilic OMPs (i.e., molinate, propanil, and dimethoate; logS > -2.5) also showed high removal efficiencies during filtration, with removal rates of more than 70%. Furthermore, an abiotic study illustrated that biodegradation is the principal mechanism through which such compounds are eliminated in the infiltration zone. Moderately-hydrophobic OMPs (atrazine, simazine, isoproturon, and metolachlor; 2.5 > logS > -4) were the most persistent compounds under the examined experimental conditions, and were mainly removed by adsorption.

The presence of heavy metals (HMs) in bank filtrate degrades its quality and causes severe damage to human health. Soil column experiments were conducted to assess the feasibility of using BF to eliminate HMs during infiltration. The impact of the organic composition of the feed waters (four different water sources were examined) on the removal of selected HMs (Pb, Cu, Zn, Ni, and Si) was explored. To provide insight into the organic composition of the different types of feed water, FEEM-PARAFAC and fluorescence indices (e.g., the humification index, fluorescence index, and biological index) were used. Experiments were performed in a temperature-controlled room (30 °C) and under aerobic conditions. Among the HMs examined, Pb had the greatest tendency for adsorption onto the sand surface and effluent Pb concentrations were below the detection level (5 µg/L) for all of the feed waters. In comparison, the removal efficiency of Cu, Zn, and Ni ranged between 65% and 95%, and depended significantly on the organic concentration and composition of the raw water. Humic compounds demonstrated the ability to reduce the removal efficiencies of these HMs during the infiltration process. Humic compounds could accumulate on the surface of sand grains and thereby reduce the sorption capacity of the sand matrix. Moreover, humic compounds also reacted with the HMs to form aqueous complexes. Conversely, biodegradable matter was found to be effective in enhancing the sorption of HMs onto the sand grains. The removal of Se was enhanced when the feed water contained a higher concentration of biodegradable matter. Nevertheless, it should be noted that high a concentration of biodegradable matter in the source water would increase the probability of shifting the infiltration environment into anaerobic conditions, and consequently enhance the mobilisation of adsorbed HMs into the infiltrating water.

The mobilisation of (toxic) metal(loid)s such as Fe, Mn, and As during infiltration, and the enrichment of their concentrations in bank filtrate, are major drawbacks that restrict the widespread implementation of BF in developing countries. DOM is the principal factor influencing the redox reactions taking place in the infiltration zone. In this research, column experiments were conducted under anaerobic conditions to assess the effect of the organic composition of the feed water on the release of Fe, Mn, and As during BF. Columns were filled with iron-oxide-coated sand and fed with different types of source water. Excitation-emission spectroscopy coupled with a parallel factor framework clustering analysis model was used to measure the organic characteristics of the feed water and fluorescence indices were also estimated. The concentrations of Fe, Mn, and As in the filtrate were in the range 10–20 µg/L, 1,500–3,900 µg/L, and < 2–7.1 µg/L, respectively, indicating that Mn-reduction was the prevailing mechanism under the experimental conditions. The mobilisation of such metal(loid)s was significantly dependent on the organic composition and concentration of the feed water. Humic compounds were observed to have a positive effect on the release of the metal(loid)s from the soil into the infiltrating water. Humic compounds have a high shuttle-electron capacity and can, therefore, function as chelators and react with the geogenic metal(loids) to form aqueous complexes. Moreover, these compounds may function as mediators of the microbial reduction of these metal(loids) in the infiltration region. The fluorescence results showed that humic substances—regardless of their origin—have adequate capability to release Mn into the infiltrating water. Furthermore, terrestrial humic compounds (condensed structure) exhibited higher efficiency in mobilising Fe. Additionally, batch studies were conducted to assess the impact of humic, fulvic, and tyrosine concentrations in the mobilisation of metal(loid)s. The experimental results showed that humic and fulvic compounds at low concentrations (≤ 5 mg-C/L) had the same capability to release Mn. The same trend was observed for Fe mobilisation. However, fulvic compounds (i.e., lower molecular weight humic compounds) at higher concentrations demonstrated a greater ability to mobilise Mn. Humic compounds, on the other hand, were comparatively more effective than fulvic compounds at mobilising Fe at higher concentrations of humics. However, these findings require further verification under different soil conditions and with different metal(loid) chemical compositions and structures. Biodegradable organic matter was also identified as being effective at mobilising Fe and Mn in the infiltration zone. Batch studies revealed that biodegradable matter (i.e., tyrosine) at concentrations greater than 10 mg-C/L was capable of creating adequate to create a Fe-reducing environment in the infiltration region, and thereby, enriching the concentration of Fe in the filtrate water.

The effective implementation of BF is highly dependent on the surrounding environmental and hydrological conditions. Here, the feasibility of implementing BF in an arid environment (Aswan, Egypt) was investigated. The overarching objective was to establish guidelines for the management of BF systems under these environmental conditions. A multidisciplinary approach was taken in order to achieve this objective, which included the following activities: (i) the development of a hydrological model to determine the impact of environmental variables on the performance of BF; (ii) a water quality monitoring study to characterise the quality of the bank filtrate; and (iii) an economic viability analysis to compare BF with existing treatment techniques. From a hydrological perspective, BF is favourable under the local environmental conditions in Aswan City. However, the design parameters (e.g., number of wells, production capacity, and distance between wells) should be identified based on minimising the contribution of contaminated groundwater to the final total pumped water. In the future, the water level of the River Nile is projected to decline owing either to the construction of the Renaissance Dam in Grand Ethiopia and/or as a consequence of climate change. The model results revealed that a fall in the level of the River Nile by 0.5–1.5 m would have a substantial effect on BF design parameters (e.g., travel time, drawdown and the share of bank filtrate) in the onset operation of the wells. Nevertheless, the continued operation of BF wells for long periods (approximately 100 days) could mitigate these effects. The water quality study demonstrated that BF in Aswan City could be an effective technique to eliminate chemical contaminants and ensure good quality drinking water. However, there was an increase in terrestrially-derived humic compounds in the pumped water, which may be due to the dissolution of these compounds from the soil into the infiltrating water or the result of mixing infiltrated water with polluted groundwater. If post-chlorination is implemented, this increase in humic compounds in the pumped water may increase the potential for the formation of trihalomethanes. Finally, the Net Present Value (NPV) and the Payback Period (PBP) were used as economic indicators to determine the viability of BF relative to other treatment techniques. Results indicate that BF has a lower NPV and PBP, indicating that it is economically viable in the case study in Aswan City, Egypt. .

Overall, this study revealed that BF is an efficient and affordable technique capable of providing good quality drinking water that meets local standards in arid environments. However, the following factors should be taken into account during the design and development of BF well fields:

- Humic compounds are subject to enrichment during the infiltration process, especially at high temperatures. Therefore, post-treatment should focus on eliminating these compounds. Conversely, biodegradable material can be readily eliminated in the infiltration region;
- The removal efficiencies of OMPs with moderate hydrophobicity (approximately -2.5 > logS > -4) is low in the infiltration area, requiring long travel times (> 30 days) for efficient elimination even at infiltration temperature of 30 °C.
- BF can be rendered more viable by positioning fields close to surface water sources with low organic content. DOM (specifically humic compounds) reduces the HM adsorption efficiency and promotes the release of undesirable metals (e.g., Fe, Mn, and As) from the soil into the infiltrating water;
- BF design parameters (e.g., the number of wells and well distance/spacing) should be configured to minimise the proportion of polluted groundwater in the overall pumped water;
- Further research is required to address the issue of well clogging, which restricts the application of BF in arid countries.

SAMENVATTING

Vervuiling van het oppervlaktewater is een wereldwijd probleem. Conventionele waterzuiveringstechnieken zijn over het algemeen niet in staat de grote verscheidenheid aan verontreinigende stoffen uit het oppervlaktewater te verwijderen. Natuurlijke zuiveringstechnieken, zoals oeverfiltratie (BF), kunnen een kosteneffectief en duurzaam alternatief vormen voor geavanceerde zuiveringstechnologieën. Oeverfiltratie is een toepassing waarbij (er continu) oppervlaktewater wordt gepompt in een nabije put, zodat er een continue watertoevoer ontstaat naar een (de) ondergrondse filtrerende watervoerende laag. De ondergrondse sedimenten werken als natuurlijke filters die verschillende soorten vervuilende stoffen uit het water kunnen verwijderen. Deze techniek heeft zijn werkzaamheid in ontwikkelde en vochtige landen al meer dan honderd jaar bewezen en resulteert in drinkwater van hoge kwaliteit. De milieuomstandigheden kunnen de komende jaren echter door klimaatverandering worden aangetast in verschillende landen. Als gevolg hiervan hebben verschillende landen (bijvoorbeeld Egypte en India) de laatste tijd meer aandacht besteed aan oeverfiltratie om aan de vraag naar drinkwater te voldoen. De toepasbaarheid van oeverfiltratie onder deze drogere klimaatomstandigheden moet nog worden onderzocht. Landen met een aride klimaat hebben de volgende kenmerken: (i) zeer variabele hydrologische omstandigheden en matige oppervlaktewaterinfiltratie in de aanliggende watervoerende laag. Als gevolg hiervan zal de reistijd van het geïnfiltreerde water naar verwachting langer worden en zal kans op anaerobe omstandigheden toenemen. (ii) Een hoge temperatuur, waardoor biologische activiteit tijdens de infiltratie hoger is en de oeverfiltratiebehandelingsmechanismen worden beïnvloedt. (iii) Het oppervlaktewater is sterk vervuild met vooral organische stoffen die de reacties in de infiltratiezone aanzienlijk beïnvloeden. Oeverfiltratie is een plaatsgebonden techniek, en er is een duidelijk tekort aan kennis over het ontwerp en het beheer van deze duurzame techniek in extreme omstandigheden met een droog klimaat en sterk vervuild oppervlaktewater. Het doel van deze thesis was om de werkzaamheid van oeverfiltratie te analyseren in dergelijke milieuomstandigheden, waarbij het verwijderen van de chemische verontreinigingen en het leveren van drinkwater van goede kwaliteit -dat voldoet aan de lokale normen- centraal staat.

Opgelost organisch materiaal (DOM) wordt beschouwd als een drijvende factor voor fysische processen en biochemische reacties die voorkomen in het infiltratiegebied en staat centraal als het gaat om kwaliteit van oeverfiltratie. Dit onderzoek is gericht op het volgen van het gedrag van organisch materiaalfracties tijdens de filtratie met behulp van geavanceerde analysetechnieken zoals: grootte-selectieve vloeistofchromatografie met organische koolstofdetectie en organische stikstofdetectie (LC-OCD/OND) en matrix excitatie-emissie fluorescentiespectroscopie met parallelle factoranalyse (FEEM-PARAFAC). Op laboratoriumschaal zijn batch- en bodemkolomexperimenten uitgevoerd om de effecten van milieuparameters (temperatuur (20-30 °C) en redoxcondities) op de aantasting van organische bestanddelen in het filtratiegebied te onderzoeken. Het experiment werd uitgevoerd met behulp van vier soorten water met verschillende herkomst. De experimentele resultaten toonden aan dat de groep van non-humische verbindingen (bijvoorbeeld biopolymeren) het meest vatbaar is voor afbraak tijdens infiltratie. Ongeacht de omgevingstemperatuur bedraagt de afbraak van biopolymeren meer dan 80% van het totaal onder zuurstofrijke omstandigheden. Omgekeerd vertonen humusverbindingen een temperatuurafhankelijk gedrag waarbij een lagere temperatuur (20-25 °C) optimaal is. Bij een hoge temperatuur (30 °C) werd echter een toename van humusverbindingen in het effluent waargenomen. Dit kan worden toegeschreven aan (i) het oplossen van bodem-organisch materiaal in het geïnfiltreerde water en/of (ii) de aanwezigheid van micro-organismen die in staat zijn labiele verbindingen om te zetten in sterkere verbindingen. Uit de analyse van de organische fluorescentie is gebleken dat er een toename is van de humusverbindingen van sedimentaire oorsprong in het effluent bij hoge temperatuur.

Oeverfiltratie heeft aangetoond dat er in ontwikkelde landen een relatief hoog potentieel is om verbindingen van organische microverontreinigingen (OMP's) te verwijderen. Deze werking is echter nog nooit voldoende onderzocht in een droog milieu. In dit onderzoek is op laboratoriumschaal een batchstudie uitgevoerd om het gedrag van verschillende OMP's in het filtratiegebied te onderzoeken. De filtratie efficiëntie van verschillende klassen OMP's (polyaromatische koolwaterstoffen, insecticiden en pesticiden) werden onderzocht voor de verblijftijd van 30 dagen onder wisselende omstandigheden (temperatuur 20-30 °C in oxische, anoxische en anaerobe condities) in onbehandeld water. Uit de resultaten bleek dat de verwijdering van OMP's en hun reacties op de omringende milieuomstandigheden in hoge mate afhankelijk zijn van de chemische eigenschappen van de OMP's (o.a. oplosbaarheid). Hydrofobe OMP's (zoals DDT, pyriproxyfen, pendimethalin, β-BHC, endosulfan sulfaat en PAK's; logS < -4) hebben de neiging om te adsorberen (>80%) aan zandkorrels, ongeacht de temperatuur en de redoxcondities. Hydrofiele OMP's (o.a. molinaat, propanil en dimethoaat, logS > -2,5) volgen dezelfde trend en werden effectief verwijderd tijdens het filtratieproces (> 70%). De abiotische studie toonde aan dat biologische degradatie het belangrijkste mechanisme is om dergelijke verbindingen in de infiltratiezone te verwijderen. Matig hydrofobe OMP's (atrazine, simazine, isoproturon en metolachloor, 2.5 > logS > -4) hadden de meest persistente eigenschappen in de onderzochte experimentele omstandigheden en werden voornamelijk verwijderd door middel van adsorptie.

De aanwezigheid van zware metalen (HM) in het oeverfiltraat vermindert de kwaliteit ervan en veroorzaakt ernstige schade aan de menselijke gezondheid. Bodemkolomexperimenten werden uitgevoerd om te onderzoeken of oeverfiltratie geschikt is om zware metalen te filtreren. De impact van de organische samenstelling (in de vier soorten water met verschillende herkomst) op de verwijdering van geselecteerde zware metalen (Pb, Cu, Zn, Ni en Si) werd onderzocht in dit onderzoek. De organische samenstelling van het water is bepaald met FEEM-PARAFAC en fluorescentie-indices (bijv. humificatie-index, fluorescentie-index en biologische index). De experimenten werden uitgevoerd in een kamer met een gecontroleerde temperatuur van 30 °C in een aerobe omgeving. Van de onderzochte zware metalen adsorbeerde lood (Pb) het best aan de zandmatrix; de loodconcentratie van het effluent lag onder het detectieniveau (5 µg/L) voor alle watergroepen. Ter vergelijking: de verwijdering van koper, zink en nikkel varieerde tussen 65 en 95%; en was sterk afhankelijk van de organische stof concentratie en -samenstelling in het ruwe water. Humusverbindingen blijken geschikt om het filterrendement van zware metalen te verbeteren. Humusverbindingen kunnen zich ophopen op het zandoppervlak en op deze manier de sorptiecapaciteit van het zand verminderen. Daarnaast kunnen humusverbindingen zelfs reageren met de zware metalen en kunnen er complexen worden gevormd. Daarentegen is gebleken dat biologisch afbreekbaar organisch materiaal de sorptie van zware metalen aan de zandkorrels kan verbeteren. De Se-verwijdering werd verbeterd wanneer het voedingswater een hogere concentratie biologisch afbreekbare stoffen bevatte. Toch moet worden opgemerkt dat een hoge concentratie van biologisch afbreekbare stoffen in het bronwater de kans vergroot op een anaeroob infiltratiemilieu waardoor eerder geadsorbeerde zware metalen worden gemobiliseerd en terecht komen in het water.

De mobilisatie van toxische metalen (zoals Fe, Mn, As) tijdens de infiltratie waardoor concentraties van deze metalen in het oeverfiltraat toenemen is een belangrijk nadeel dat de toepassing van de oeverfiltratietechniek beperkt, vooral in ontwikkelingslanden. Opgelost organisch materiaal is de belangrijkste factor die de redoxreacties in de infiltratiezone beïnvloedt. In dit onderzoek werd een anaeroob kolomexperiment uitgevoerd om de rol van de organische samenstelling van het water te onderzoeken bij het vrijkomen van ijzer, mangaan en arsenicum tijdens oeverfiltratie. De kolommen werden gevuld met zand dat gecoat was met ijzeroxide waarna verschillende soorten water werden geïnfiltreerd. De excitatie-emissie-spectroscopie technieken gekoppeld aan het parallelle factor-framework-cluster-analysemodel werden gebruikt om de organische kenmerken van de waterbronnen in kaart te brengen. Ook werden de fluorescentie-indices geschat. De concentraties van Fe, Mn en As in het filtraat varieerden respectievelijk tussen 10-20, 1500-3900 en <2-7,1 µg/L. Dit geeft aan dat de mangaanreductie het belangrijkste mechanisme was onder de onderzochte experimentele omstandigheden. De mobilisatie van dergelijke metalen was sterk afhankelijk van de organische samenstelling en concentratie in het voedingswater. Humusverbindingen bleken een positief effect te hebben op het vrijkomen van de metalen uit de bodem. Humusverbindingen hebben een hoge shuttle-elektron capaciteit en kunnen dus functioneren als chelaten en reageren met de geogene metalen zodat complexen worden gevormd. Bovendien kunnen ze bijdragen aan de microbiële reductie van deze metalen in het infiltratiegebied. De fluorescentieresultaten toonden aan dat humusverbindingen, ongeacht hun herkomst, in staat zijn om mangaan te mobiliseren zodat het in het infiltrerende water terecht komt. Daarnaast zorgden aardse humusverbindingen (gecondenseerde structuur) er voor dat er meer ijzer gemobiliseerd werd. Ook zijn er in dit onderzoek batchstudies uitgevoerd om de impact van humus-, fulvine- en tyrosineconcentraties op de mobilisatie van metalen te onderzoeken. De experimentele resultaten toonden aan dat humus- en fulvineverbindingen bij een lage concentratie ($\leq$ 5 mg-C/L) hetzelfde vermogen hadden om mangaan te mobiliseren. Dezelfde trend was waargenomen voor ijzermobilisatie. Echter waren de fulvineverbindingen (humusverbindingen met een lager moleculair gewicht) in hogere concentraties beter in staat om mangaan te mobiliseren. Humusverbindingen waren in hogere concentraties daarentegen effectiever dan fulvineverbindingen om ijzer te mobiliseren. Deze bevindingen moeten echter nog worden onderzocht in verschillende bodemsoorten met verschillende chemische metaalsamenstellingen en -structuren. Biologisch afbreekbare organische stoffen bleken ook ijzer en mangaan te mobiliseren in de infiltratiezone. Uit het batchonderzoek is gebleken dat biologisch afbreekbare stoffen (bv. tyrosine) bij een concentratie van meer dan 10 mg-C/L voldoende zijn om een ijzer-reducerend milieu in het infiltratiegebied te creëren zodat de ijzerconcentratie in het filtraat toeneemt.

De toepassing van oeverfiltratie is sterk afhankelijk van de milieu- en hydrologische omstandigheden. De haalbaarheid van oeverfiltratie in een droge omgeving (Aswan, Egypte) is in deze studie onderzocht. Het overkoepelende doel van deze studie was om richtlijnen op te stellen voor het beheer van oeverfiltratiesystemen onder deze milieuomstandigheden. Om dit doel te bereiken werd een onderzoek uitgevoerd waarin drie disciplines zijn samengekomen, waaronder (i) de ontwikkeling van een hydrologisch model om de impact van milieuvariabelen op het resultaat van oeverfiltratie te bepalen. (ii) Er is een waterkwaliteitsonderzoek uitgevoerd om de kwaliteit van het oeverfiltraat te bepalen. iii) Er is een economische rendabiliteitsanalyse uitgevoerd om oeverfiltratie te vergelijken met bestaande zuiveringstechnieken. Vanuit hydrologisch oogpunt is oeverfiltratie een gunstige techniek onder de lokale milieuomstandigheden van Aswan City. De ontwerpparameters (bv. aantal putten, productiecapaciteit, afstand tussen de putten) moeten echter vooral worden bepaald op basis van minimalisatie van het oppompen van verontreinigd grondwater. Het waterpeil van de rivier de Nijl zal naar verwachting in de nabije toekomst dalen als gevolg van de bouw van de Renaissancedam in Grand Ethiopië of ten gevolge van klimaatverandering. De modelresultaten hebben aangetoond dat de daling van het peil van de Nijl (met 0,5-1,5 m) een substantieel effect heeft op de oeverfiltratieparameters (bv. reistijd, het aandeel van het oeverfiltraat) bij het in werking stellen van de putten. Toch zouden oeverfiltratieputten gedurende een lange periode (ongeveer 100 dagen) de gevolgen kunnen verzachten. Het waterkwaliteitsonderzoek toonde aan dat oeverfiltratie in Aswan City een effectieve techniek is om de chemische verontreinigingen te elimineren en een behoorlijke kwaliteit van het drinkwater te garanderen. Er was echter een toename van de humusverbindingen (terrestrisch) in het opgepompte water, welke het gevolg kan zijn van het oplossen van deze verbindingen uit de bodem in het infiltrerende water of van het mixen met het verontreinigde grondwater. Als er chloor wordt toegevoegd aan het gefilterde water kan de toename van humusverbindingen de kans op vorming van trihalogeenmethaanverbindingen vergroten. Uiteindelijk werden de kapitaalwaarde (NPV) en de terugverdientijd (PBP) gebruikt als economische karakteristieken gebruikt om de levensvatbaarheid van oeverfiltratie te bepalen ten opzichte van andere zuiveringstechnieken. De lage berekende NPV- en PBP-waarden betekenen dat de oeverfiltratietechniek een gezonde economische levensvatbaarheid heeft.

Samengevat heeft deze studie aangetoond dat oeverfiltratie een efficiënte en betaalbare techniek is om drinkwater van goede kwaliteit te leveren, dat voldoet aan de lokale normen in landen met een droog milieu. Tijdens de ontwerp- en ontwikkelingsfasen moet echter rekening worden gehouden met een aantal punten:

- Tijdens het filtratieproces kan de concentratie humusverbindingen toenemen, vooral bij hoge temperaturen. Daarom moet nabehandeling voornamelijk gericht zijn op het elimineren van deze verbindingen. Wel kan biologisch afbreekbaar materiaal gemakkelijk worden verwijderd tijdens filtratie.

- Matig hydrofobe OMP's (ruwweg -2,5>logS>-4) worden nauwelijks verwijderd in het infiltratiegebied en er is een lange reistijd (>30 dagen) nodig om dergelijke verbindingen te verwijderen.

- Het gebruik van de oeverfiltratie is rendabeler wanneer oppervlaktewater wordt gebruikt met een laag gehalte organische stof. Opgeloste organische stof (DOM, specifiek humusverbindingen) vermindert de adsorptie efficiëntie van zware metalen en bevordert het vrijkomen van ongewenste metalen (bijv. Fe, Mn en As) uit de bodem in het infiltrerende water.

- De oeverfiltratie-ontwerpparameters (bv. aantal putten en putafstand) moeten zo worden bepaald dat het aandeel van verontreinigd grondwater in het totale opgepompte water tot een minimum wordt beperkt.

Verder onderzoek moet worden verricht om andere problemen aan te pakken, zoals verstoppingsprocessen, die de toepassing van oeverfiltratie in droge landen in de weg staan.

CONTENTS

1
INTRODUCTION

1.1 BACKGROUND

The demand for high-quality drinking water is growing dramatically throughout the world, particularly with a rise in urbanisation and population growth. However, contamination of surface water resources through the discharge of municipal and industrial wastewaters necessitates intensive industrial treatment (Ray et al., 2011). Schwarzenbach et al. (2006) reported that chemical contamination of surface water is a serious environmental problem facing humanity. In the last few decades, a growing variety of environmental contaminants have been detected at elevated concentrations in freshwater resources including organic compounds such as humic acids, compounds used in personal care products, pesticides, insecticides, pharmaceuticals, and synthetic chemicals; and inorganic chemicals such as nitrogen, phosphorus, and metals. This presents numerous challenges for drinking water treatment systems, such as odour, colour, and taste, as well as raising the required dosage of chemicals for coagulation and disinfection processes (Matilainen et al., 2010). Most developing countries employ conventional water-treatment technologies that are no longer considered viable for effectively disinfecting polluted water and eliminating or reducing contaminants to the required levels (Maeng, 2010).

Arid and semi-arid societies face even more severe water management challenges due to the scarcity of water resources (Sophocleous, 1997, 2002), Indeed, many countries with an arid climate struggle to supply good quality drinking water at a low economic cost. The hydrological conditions in arid climates can be extreme and highly variable (Sophocleous, 2000). High temperatures primarily influence the effectiveness of conventional treatment processes, such as adsorption, coagulation and disinfection and thereby, the quality of drinking water provided (Sugiyama et al., 2013). Moreover, this increases the required chemical dosages for disinfection processes (LeChevallier, 2004). Geriesh et al. (2008) suggested that pre-treatment involving filtering surface water and reducing organic content would increase the quality of drinking water supplied in arid climates. Therefore, natural treatment systems, such as bank filtration (BF), offer potentially viable options for water supply schemes in arid and semi-arid areas. These systems involve treatment and/or pre-treatment steps to remove pathogens, algal toxins and organic matter (OM), and reduce turbidity and chemical pollutants in the drinking water (Hiscock et al., 2002).

1.2 BANK FILTRATION

BF is regarded as a simple and sustainable technique that can provide good -quality drinking water. After the Second World War, European surface water resources became heavily contaminated with industrial and municipal waste water, and BF was the only means considered able to secure drinking water of acceptable quality (Shamrukh et al., 2011). BF is a process in which surface water undergoes a subsurface flow caused by the lowering of the hydraulic head prior to abstraction from vertical or horizontal wells (Grischek et al., 2003). The raw water extracted from the production well consists of a mixture of infiltrated surface water and ambient groundwater. It has been shown that under suitable hydrogeological conditions, well-operated BF facilities may provide relatively low-cost, high-quality drinking water that requires little further treatment (Tufenkji et al., 2002). Alluvial aquifers are the most suitable sites given their high production capacity, high connectivity to surface water sources, and accessibility to regions of demand (Doussan et al., 1997).

BF can improve water quality effectively by reducing turbidity, microbial contaminants, microcystins, pathogens, heavy metals (HMs), OM, and inorganic pollutants (Gandy et al., 2007; Hiscock et al., 2002; Sontheimer, 1980). BF has a high capability to eliminate such soluble contaminants that are difficult to remove in surface water treatment plants. For example, BF has been shown to reduce dissolved organic matter (DOM) and disinfected by-product (DBP) precursors by 50% (Ray et al., 2002). Attenuation of pollutants relies on mechanisms such as biodegradation, adsorption, precipitation, and filtration. The effectiveness of this approach depends on a variety of factors, including aquifer geology, aquifer structure, surface water flow, surface and groundwater OM type, river bed composition and clogging processes as well as land use in the local catchment region (Hiscock et al., 2002).

BF has long been used as a multi-objective natural treatment technology that eliminates much of the surface water contamination. BF also equilibrates temperature and dampens accidental chemical load peaks. It can be used to replace or support existing water treatment techniques by providing a robust barrier and reducing the cost of treatment. BF also helps reduce the use of chemical disinfectants to produce biologically-stable water (Sharma et al., 2009). Another advantage of BF is that it may be used in regions with seasonally variable precipitation and run-off regimes (e.g., monsoon-, flood-, and drought-prone regions) as a means of increasing water-storage capacity (Hülshoff et al., 2009). Moreover, mixing bank filtrate with local groundwater increases the groundwater supply and dilutes contaminants (Grischek et al., 2003).

1.3 EXPERIENCE OF BANK FILTRATION

BF has been utilised by several water supply companies in Europe and North America for the production of drinking water. In Germany, BF is primarily used around the Rhine

3

River at the Lower Rhine in the region between the Sieg and Ruhr tributaries, the Elbe River between Dresden and Torgau, and the Berlin district. Its application also covers the Rhine and Meuse Rivers in the Netherlands, as well as in many other European countries including Austria, Switzerland, and France. BF provides 50% of France's drinking water supply, 15% in Germany, and 12% in the Netherlands (Deyi, 2012; Hiscock et al., 2002). On the Rhine River at Dusseldorf, Germany, BF has been used widely since 1870 to provide high-quality drinking water. In 1960, the quality of the river Rhine started to deteriorate and anaerobic conditions developed in the infiltration zone. Iron and manganese reduction rates also increased. As a result, there was a need for post-treatment of river bank-filtrate. In the last few decades, the water quality of the raw water has improved; however, periodic changes in river water quality and hydraulics due to climatic conditions are on-going issues (Eckert et al., 2006). Currently, granular activated carbon is used in conjunction with ozonation and filtration to further treat bank filtrate and eliminate chemical contaminants.

Horizontal collector wells have been used in the United States for more than 80 years to pump bank filtrate. Due to the high production capacity, horizontal wells were mainly used to supply water for industrial uses, although approximately one-third of the horizontal wells that were initially produced were used for drinking purposes. In recent years, this technique has been implemented to produce drinking water in vast amounts and to ensure water quality standards are met (Hunt, 2003). BF has proved effective in the removal of chemical and biological contaminants including organic micropollutants (OMPs), Giardia and Cryptosporidium parasites, and microcystins, which are not adequately removed by conventional treatment techniques (Ray et al., 2002). In North America, BF is widely used as a pre-treatment system to enhance the quality of raw water and reduce the cost of treatment (Wang, 2003).

The use of BF has expanded in developing countries in recent years, including Kenya, Malawi, Bosnia, Russia, Egypt, India, Korea, and China (Bartak et al., 2014; Bosuben, 2007; Chaweza, 2006; Dash et al., 2008; Ray et al., 2011). Most of the BF wells established in these countries are vertical wells, which are mostly recharged from the local surface water system and are not designed as BF wells (Shamrukh et al., 2011). Shamrukh et al. (2008) illustrated that mixing bank filtrate with highly-polluted groundwater, and the dissolution of iron and manganese along the infiltration path, are the main problems currently restricting the wider use of this technique in developing countries.

1.4 BANK FILTRATE QUALITY

The effectiveness of BF in the production of high-quality drinking water is dependent on a multiple of variables, including raw water quality, hydrological characteristics, and geological setting. Hydrological characteristics have substantial effects on the travel time and redox conditions of the infiltration zone, which have direct influences on BF efficiency and pumped water quality (Ray et al., 2002). This section outlines the main factors impacting the quality of bank filtrate.

1.4.1 Raw water quality

The concentration of pollutants in the raw water system is one of the main parameters affecting the performance of BF and the need for post-treatment. Schijven et al. (2003) reported that if the concentration of *Cryptosporidium* is greater than 0.075 oocysts/L in raw water, further treatment would be needed to provide safe drinking water. Kedziorek et al. (2009) stated that if the electron trapping capacity (ETC) (calculated from the summation of the dissolved O_2 and NO_3^- concentrations) of the infiltrate water is greater than 0.2 mmol/L, the concentration of manganese in the abstracted bank filtrate would be very low (< 0.1 µm) unless the ambient groundwater has a higher concentration of this contaminant. The quality characteristics of raw water are influenced by hydrological and climatic conditions. Surface water systems with low flow velocities and high nutrient concentrations have a higher potential for the formation of algal blooms. In the same regard, climate has a significant impact on redox processes taking place in surface water bodies. Furthermore, dissolution of metals from the bank and bottom sediments is high under arid conditions. For example, Zwolsman et al. (2008) observed that the concentrations of HMs (cadmium, chromium, mercury, lead, copper, nickel, and zinc) in the Rhine River were higher during the 2003 drought.

1.4.2 Travel time

Travel time has a significant effect on the efficiency of BF and should, therefore, be taken into consideration throughout the design phase (Sprenger et al., 2011). Long travel times provide more opportunity and time for sorption and biodegradation, which are essential for the elimination of chemical and biological pollutants. However, it may also have an adverse effect by enabling the development of anaerobic conditions, which can increase the dissolved metal load in the pumped water. The travel times required in the BF systems is mainly determined by the persistence of pathogens. It was suggested, for example, that travel times of 60 days in Germany and 70 days in the Netherlands are adequate to ensure biologically stable drinking water (Azadpour-Keeley, 2003; Hiscock et al., 2002). The proposed travel time to remove 90% of OMPs is > 6 months (Drewes et al., 2003). In North America, BF is often used/considered as a pre-treatment technique for conventional drinking water treatment plants, for which the travel times range from hours to a few weeks at most (Grünheid et al., 2005).

5

Various methods have been proposed to calculate travel times. Wang et al. (2008) used bromide and tritium to determine travel times and the influence of soil type and land cover on recharge rates along Hebei Plain (China). Atrazine has also been used to estimate the travel time and river discharge at the banks of the Platte River (Nebraska, United States) (Duncan et al., 1991; Verstraeten et al., 1999). Fluctuations in electrical conductivity were used to determine the travel time of infiltrated water in aquifer adjacent to the Thur River in Switzerland (Vogt et al., 2010). Dyes and temperature variations have also used as tracers to estimate travel times (Anderson, 2005; Hoehn et al., 2006; Verstraeten et al., 1999). Recently, a wide range of chemical isotopes has been used to assess infiltration processes and retention times in BF systems (Kármán et al., 2013; Vogt et al., 2010). Modelling is also applied to estimate the travel times and flow path of BF systems. For example, advection models, such as PMPATH (Bosuben, 2007), can be used to calculate pore-water velocities for estimating travel times. These studies indicate that travel time is mainly affected by the distance to the riverbank, pumping rate, drawdown, and well spacing.

1.4.3 Redox process

The nature of the redox environment is very important in BF processes as this influences the occurrence of HMs in the bank filtrate such as copper, zinc, lead, iron, and manganese (Massmann et al., 2008). Furthermore, redox conditions determine the fate of OMPs in the zone of infiltration (Maeng, 2010) and influence pH of the bank filtrate, and consequently, affect the overall BF process (Massmann et al., 2008). The biochemical processes taking place in the infiltration zone, such as the degradation of OM, denitrification, and dissolution of iron and manganese, are mainly dependent on the redox environment. Microorganisms degrade OM to produce energy using various acceptor electron species (i.e., O_2, NO_3^-, Mn^{+4}, Fe^{+3}, and SO_4^{-2}). As a consequence, reduced species are produced (i.e., H_2O, N_2, Mn^{+2}, Fe^{+2}, and HS^-) that have adverse effects on the quality of the bank filtrate (Kedziorek et al., 2009). Toxic contaminants, such as arsenate ions, can also be introduced into bank filtrate water as a consequence of redox reactions in the infiltration zone. Arsenate ion concentrations increase in bank filtrate water through the oxidation of As-bearing sulphide minerals such as arsenopyrite. Moreover, soluble arsenate complexes adsorbed onto the surfaces of iron and manganese mineral surfaces in the aquifer can be mobilised under iron- and manganese-reducing conditions, and introduced into the bank filtrate. Mobilisation of As occurs primarily in alluvial aquifers mainly composed of sand and gravels (McMahon et al., 2008).

Some redox processes are driven by specific bacterial communities. Dissolved oxygen supplemented by nitrate is used preferentially by subsurface microorganisms as they provide the most energy per mole of organic carbon oxidised than any other widely usable electron acceptor. Under anaerobic conditions (i.e., in the absence of oxygen and nitrate), the reduced forms of manganese and iron can be released from the aquifer materials into the bank filtrate (Table 1.1). Therefore, various redox zones can be discriminated based on the concentrations of biodegradable OM and electron acceptor species including oxic, anoxic (nitrate-reducing), and anaerobic (Fe and Mn reducing) zones. Several studies have distinguished these redox zones in the infiltration zone (Barcelona et al., 1992; Berner, 1981; Champ et al., 1979; Heron et al., 1993). A framework for distinguishing the redox zones was proposed by Kedziorek et al. (2009), who stated that the threshold concentrations indicative of denitrification, manganese, iron, and sulphate reduction are 0.5 mg/L (NO_3^-), 0.05 mg/L (Mn^{+2}), 0.1 mg/L (Fe^{+2}), and 0.5 mg/L (SO_4^{-2}), respectively (Table 1.1). The denitrification process takes place when the concentration of dissolved oxygen is less than 0.5 mg/L. In contrast, Seitzinger et al. (2006) stated that the onset of denitrification requires dissolved oxygen concentrations between 0.2 mg/L and 0.3 mg/L. McMahon et al. (2008) pointed out that redox reactions are not only associated with the availability of electron acceptors but are also influenced by the hydrological and climatic conditions of the aquifer as well as soil grain size, composition, organic content, the microbial species present, and temperature (Lynch et al., 2014).

The area and activity of each redox zone is primarily affected by climatic conditions. Gross-Wittke et al. (2010) specified that all redox processes are influenced by air and water temperatures. Although the prevailing redox state in most BF sites in Switzerland is oxic, the rise in water temperature in the summer of 2003 (by 3.5 °C above the average annual temperature) resulted in anaerobic conditions that raised the concentrations of iron and manganese in the pumped water (Rohr, 2014). Gross-Wittke et al. (2010) studied the influence of temperature (5–25 °C) on redox processes. They concluded that increased temperatures reduce oxygen solubility in the raw water and intensify biological activity in the infiltration zone, leading to increased oxygen consumption and decreased redox potential in the sediments. Massmann et al. (2008) studied the redox conditions in an aquifer adjacent to Lake Wannsee in Berlin, Germany. The authors concluded that redox conditions display a strong seasonality as a result of variations in microbial activity driven by significant temperature variations in the lake of almost 25 °C. The temperature projections issued by (Stocker et al., 2013) indicate a rise of 6.5 °C by 2100, which would have a significant influence on bank filtration across the world.

Table 1.1. Threshold concentrations for identifying redox processes in the aquifer system (McMahon et al., 2008)

Redox Process	Water Quality Criteria (mg/L)				
	O_2	NO_3^-	Mn^{+2}	Fe^{+2}	SO_4^{-2}
Oxic					
O_2 reduction	≥ 0.5	----	<0.05	<0.1	----
Anoxic					
NO_3^- reduction	<0.5	≥ 0.5	<0.05	<0.1	----
Mn^{+4} reduction	<0.5	<0.5	≥ 0.05	<0.1	----
Fe^{+3}/SO_4^{-2} reduction	<0.5	<0.5	----	≥ 0.1	≥ 0.5
Methanogenesis	<0.5	<0.5	----	≥ 0.1	<0.5

1.5 IMPACTS OF CLIMATE ON BANK FILTRATION

BF processes are particularly sensitive to local and regional climatic conditions, yet the effects of climate on the quantity and quality of bank filtrate are not fully understood. Temperature and precipitation are the key climatic variables influencing water availability and BF performance (Derx et al., 2012). Sprenger et al. (2011) demonstrated that climate has both direct and indirect effects on BF performance (Figure 1.1); climate affects raw water quality, river water level, the river discharge and run-off regimes that determine infiltration rates, travel times, and the aquifer redox environment, all of which

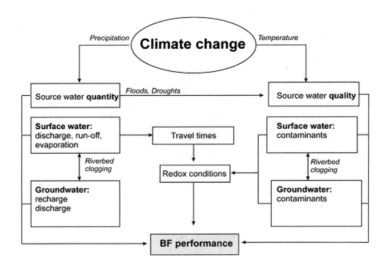

Figure 1.1. Effects of climate on BF performance (Sprenger et al., 2011).

are important variables affecting BF performance and the elimination of potential contaminants.

1.5.1 Impact of climate on raw water quality

The quality characteristics of surface water sources are substantially altered in reaction to climatic conditions. Temperature increase affects water quality variables by lowering the solubility of dissolved oxygen, enhancing the rates of biochemical reactions, modifying the stratification patterns in the surface water system, and indirectly, by increasing evaporation rate, decreasing the quantity of surface water and hence, increasing the concentrations of pollutants (Verweij et al., 2010). In the hot summer of 2003, the level of the Rhine River fell to 2 m below the average annual level (30 m above sea level [a.s.l.]), surface water temperatures approached 25 °C, and dissolved oxygen loads fell from 13 mg/L (supersaturation) to 7 mg/L (Eckert et al., 2008; Hülshoff et al., 2009). Moreover, increases in water temperatures promote the mineralisation and release of nutrients (nitrogen, phosphorus, and carbon) from soil OM in the surface water. High temperatures also facilitate the growth of toxic cyanobacteria such as *Microcystis* over diatoms and green algae (Delpla et al., 2009).

Overall, concentrations of pathogens, algal toxins, anions and cations, HMs, OM, and trace organic compounds are typically enhanced under arid conditions. Van Vliet et al. (2008) researched the effects of drought on the water quality of the Meuse River in Europe during the hot summers of 1976 and 2003, concluding that water quality was largely degraded under drought conditions (maximum temperature = 26.9 °C). Furthermore, median values of chlorophyll-*a* were higher in the hot summer of 1976 (25 µg/L) and 2003 (38 µg/L) compared to the reference years of 1978 (18 µg/L) and 2004 (12 µg/L). The concentrations of ammonia, nitrite, and orthophosphate were also higher in 2003 (0.76 mg/L, 0.20 mg/L, and 0.30 mg/L, respectively) compared to 2004 (0.46 mg/L, 0.13 mg/L, and 0.28 mg/L, respectively). Similarly, the concentrations of halogens (Cl$^-$, Br$^-$, and F$^-$), major cations and anions (SO$_4^{-2}$ and K$^+$), and trace elements (Pb, Cr, Hg, Cd, Zn, Cu, As, Ni, Ba, and Se) were higher during the drought years (1976 and 2003) relative to the reference years (1977 and 2004). These findings suggest that drought followed by low river discharge rates have adverse effects on water quality, especially in the presence of pollution sources. On the other hand, suspended solid content has been found to be marginally lower under arid conditions owing to the lower transport potential of suspended solids under low-flow circumstances (Van Vliet et al., 2008), which reduces clogging of the river bed.

1.5.2 Impact of climate on travel time

The interaction between the surface water and groundwater depends considerably on seasonal climatic conditions. Higher evaporation rates and lower surface water quantity during arid conditions prolong travel times and directly impact the quality of bank filtrate. For example, during the hot summer of 2003, the river level in the BF field of the Rhine River, Dusseldorf, Germany, decreased to 28 m a.s.l. (2 m below the annual mean level), which result in a 0.5-m drawdown in the BF wells and an extraction rate of approximately 15 m^3/h. When the pumping rate was doubled, the drawdown was increased to 1.5 m (Eckert et al., 2008). Furthermore, the reduction of surface water availability and the longer travel time resulted in an increase in the ratio of ambient groundwater in the pumped water. To address the issue of a lower river level and longer travel time, in Berlin, a weir system was established to avoid temporary drops in the surface water level during periods of low flow, thereby maintaining the hydraulic head (Hülshoff et al., 2009).

1.5.3 Impacts of climate on the redox conditions

The redox process decreases significantly as temperature increases. Gross-Wittke et al. (2010) showed that during field research at BF fields along Lake Tegel (Berlin, Germany), the redox potential decreased with decreasing temperature. The pore water redox potential changed from +178 to -14 mV as the temperature increased from 16 and 20 °C. However, the redox range decreased from +17 to -47 mV as the temperature at the site under investigation rose to 21-25 °C. NO_3-concentration declines dramatically in infiltrated water with higher temperatures (5-25 °C), suggesting that microbial denitrification and ammonization processes are favourable at higher temperatures, evidenced by the observed higher ammonia concentration. In that respect, Fe^{+3} and Mn^{+4} mobilization rates increased with increasing temperature. The levels of Fe^{+2} and Mn^{+2} increased in the bank filtrate of the investigated field from ≥ 50 µg/L at 5°C to 100 µg/L Fe^{+2} and 311 µg/L Mn^{+2} at 25 °C. Extrapolation of the data suggested that the reduction rate increased at a higher temperature (30 °C) (Gross-Wittke et al., 2010). The increase in the rate of redox reactions at elevated temperatures is ascribed to the following: i) Higher water temperatures reduces the solubility of oxygen. Oxygen consumption in the surface water will enhance anaerobic conditions at the infiltration area, and thus anaerobic reduction process is expected. ii) Temperature increase stimulates the growth of algae and microbial activity, contributing to increased oxygen consumption, thus turning the BF environment from aerobic to anaerobic (Sprenger et al., 2011).

1.5.4 Impact of climate on BF quality

The quality of bank filtrate is primarily defined by the infiltration process and the quality of the raw water. For example, climate conditions can strongly influence the concentrations of pollutants in raw water and the rate of biochemical reactions occurring in the infiltration zone. BF is used as a method to eliminate OM from water (Maeng et al., 2010). Nevertheless, temperature rises and droughts could enhance the development of anaerobic conditions in the infiltration zone, which is not an effective environment for the removal of organic pollutants. Grünheid et al. (2005) reported a preferential removal of non-biodegradable OM during winter at a BF field adjacent to Lake Tegel in Germany. Likewise, Maeng (2010) reported that the removal of OMPs (e.g., phenazone-type OMPs and pharmaceuticals) is higher in winter than summer, largely due to the development of oxic conditions in the filtration zone during winter. Another explanation for lower OM elimination rates in summer is the higher rate of decomposition of soil OM at higher temperatures. In the case of drought and low soil moisture levels, organic pollutants are adsorbed onto soil particles and deposited in the soil in the solid form, which may then be leached during the flow process (Huang et al., 2013). According to Brettar et al. (2002), a large portion of bank filtrate OM comes from the soil matrices within the infiltration zone.

The inorganic characteristics of bank filtrate are also influenced by the climate of the BF region. Warm water contains less dissolved oxygen and a higher concentration of OM (Sprenger et al., 2011). Hence, denitrification and other redox reactions are elevated in arid environments and subsequently, bank filtrate could contain higher concentrations of ammonia, iron, and manganese as well as lower nitrate concentrations. Furthermore, the dissolution and desorption HMs is higher under anaerobic conditions. In the same regard, travel time might be prolonged in dry conditions, which has a detrimental effect on bank filtrate quality by enhancing dissolution (Sprenger et al., 2011).

On the other hand, in arid environments, the removal efficiency of BF may be enhanced. Arid conditions promote the development of anaerobic conditions in the infiltration region, which are more favourable for the elimination of aromatic and double bonded compounds in comparison to aerobic conditions, which are more efficient in removing aliphatic OM. For example, under aerobic conditions, specific ultraviolet absorbance (SUVA) was determined to be 2.5 mg/L in BF wells in Lake Tegel, Berlin, Germany, compared to 2.1 mg/L under anaerobic conditions (Grünheid et al., 2005). Similarly, longer travel times under anaerobic conditions in temperate climates facilitate the elimination of several OMPs (e.g., Trihalomethane (THM) disinfection-by-products) and pharmaceuticals (e.g., sulphamethoxazole and amidotrizoic acid) (Hülshoff et al., 2009). Sprenger et al. (2011) illustrated that the half-life of dichloroethene (DCE) is 39 days under aerobic conditions but just 4.06 days under anaerobic conditions. These findings demonstrate that anaerobic conditions in the infiltration zone can be conducive to the removal of specific micropollutants.

1.5.5 Impact of climate on bank filtrate yield

It is recognised that temperature influences the viscosity of water and therefore, determines the rate of infiltration. As a consequence, the level of drawdown needed for a given flow fluctuates with variations in water temperature (Derx et al., 2012). Therefore, the impact of seasonal temperature variations on the quality and quantity of bank filtrate should also be taken into account. Caldwell (2006) used the bulk specific capacity concept to examine the effectiveness of BF at various temperatures. The bulk specific capacity can be computed by dividing the total production capacity of the BF field by the average drawdown in the wells at a given temperature and hence, the maximum capacity of the BF system can be estimated by multiplying the desired specific capacity at a defined temperature by the allowable drawdown. Caldwell (2006) used this approach to examine the influence of water temperature on production capacity at four BF fields on the Ohio, Raccoon, Platte, and Missouri Rivers in the United States. It was concluded that bulk specific capacity followed the same pattern as temperature over the year, and that temperature had a marked effect on BF yields. The Pearson's r correlation between temperature and bulk specific capacity in these regions ranged between 0.5 and 0.85. Furthermore, the specific capacity of the BF field on the Raccoon River more than doubled as temperature cycled from its minimum to maximum values (2–29 °C). Therefore, Caldwell (2006) specified that temperature should be taken into account during the design, operation, and performance evaluation of BF fields.

1.6 RELEVANCE OF THE RESEARCH

BF is widely accepted as a reliable and affordable means of supplying high-quality drinking water in Europe and North America, where hydrological conditions, raw water quality, and environmental conditions are favourable (Grischek et al., 2003). Nevertheless, these countries are vulnerable to the impacts of climate change and therefore, surface and subsurface systems are particularly prone to change in the near future. Recent climate models predict a 1.4–5.8 °C increase in average global temperatures by 2099 (Misra, 2014). Such an increase will impact the quality of surface water systems and subsequently, the treatment mechanisms in the zone of infiltration (Gross-Wittke et al., 2010). Therefore, there is a need to study the potential impacts of climate change on BF performance and bank filtrate quality.

More recently, the application of BF has been extended to developing and industrialised countries (e.g., China, Egypt, and India), which have different climatic and environmental conditions (Bartak et al., 2015; Ghodeif et al., 2016). Many of these countries are characterised by arid and semi-arid conditions and highly variable hydrological conditions. Under drought conditions, surface water flow and infiltration rates to the adjacent aquifer are reduced, which in turn prolongs the travel time and impacts BF efficiency. Furthermore, surface water bodies in these regions are typically heavily contaminated with organic and inorganic pollutants. For example, 1.3 million m^3 of treated wastewater, with a high organic load,is discharged annually through 238 sewage treatment plants to the Nile River in Egypt (Shamrukh et al., 2011). The level of pollution in surface water systems might increase the potential for the development of anaerobic conditions in the infiltration zone and thereby, affect the behaviour and fate of contaminants.

The lack of experience in managing BF is an additional drawback that limits the wider application of this sustainable technique in developing countries. BF is recognised as a site-specific technique and in-depth field investigations are required to determine the feasibility of implementing BF schemes under local hydrological conditions (Bartak et al., 2014). To address these knowledge gaps, this study aims to evaluate the efficacy of BF in providing clean drinking water in arid region with heavily contaminated surface water.

1.7 RESEARCH OBJECTIVES

The overarching aim of this thesis is to examine the application of BF in developing nations with hot, arid climates. To achieve this, a better understanding of the behaviour of contaminants in the infiltration zone and how they are influenced by environmental factors (e.g., raw water characteristics and temperature) is required. As such, the specific objectives of this thesis are:

(i)   To track changes in organic matter fractions during BF under different environmental conditions;

(ii)   To analyse the impact of temperature, organic matter composition, , and redox conditions on the removal of OMPs;

(iii)   To investigate the behaviour of HMs in the infiltration zone under different environmental conditions;

(iv)   To assess the impact of environmental conditions on the mobilisation of Fe, Mn, and As during BF;

(v)   To analyse the performance of BF in an arid city (Aswan City, Egypt).

1.8 OUTLINE OF THE THESIS

The thesis is structured in seven chapters, as follows:

Background information on the use of the BF in water treatment, knowledge gaps in its application in arid climate conditions, and the specific research objectives have been outlined in this chapter (**Chapter 1**).

Chapter 2 examines the attenuation of organic matter and its constituents during BF based on laboratory-scale column and batch experiments performed under varying conditions of temperature, raw water organic composition, and redox conditions.

Chapter 3 discusses the removal efficiencies of various classes of OMPs (polyaromatic hydrocarbons, insecticides, and pesticides) with varying physical and chemical properties during BF.

Chapter 4 discusses the removal of selected heavy metals (Cu, Zn, Pb, Se, and Ni) under different environmental conditions investigated in the laboratory-scale column experiments.

Chapter 5 summarises findings on the impact of the organic composition of raw water on the release of (toxic) metal(loids), such as Fe, Mn, and As, from the soil into the infiltrated water under anaerobic conditions.

In **Chapter 6**, the effectiveness of BF in providing drinking water of suitable quality in Aswan City, Egypt, is investigated. For this, a groundwater model is developed and implemented to assess the impact of hydrological parameters on BF performance. A water quality study is described that determines the quality of bank filtrate. Finally, a cost-benefit analysis is conducted to assess the economic viability of BF in Aswan in comparison to alternative treatment technologies. .

Chapter 7 discusses the main findings of the research and proposes a set of guidelines for the effective application of BF in arid countries. The need for further research to promote the use of BF in such countries is also highlighted.

2

REMOVAL OF DISSOLVED ORGANIC MATTER DURING BANK FILTRATION

ABSTRACT

Dissolved organic matter (DOM) in source water highly influences the removal of different contaminants and dissolution of aquifer materials during bank filtration (BF). The fate of DOM during BF processes under hot arid climate conditions was analysed by conducting laboratory–scale batch and column studies under different environmental conditions with varying temperature (20-30°C), redox, and feed water organic matter composition. The behaviour of the DOM fractions was monitored using various analytical techniques: fluorescence excitation-emission matrix spectroscopy coupled with parallel factor analysis (PARAFAC-EEM), and size exclusion liquid chromatography with organic carbon detection (LC-OCD/OND). The results revealed that DOM attenuation is highly dependent ($p<0.05$) on redox conditions and temperature, with higher removal at lower temperatures and oxic conditions. Biopolymers were the fraction most amenable to removal by biodegradation (>80%) in oxic environments irrespective of temperature and feed water organic composition. The removal was 20-24% lower under sub-oxic conditions. In contrast, removal of humic compounds exhibited higher dependency on temperature. PARAFAC-EEM revealed that terrestrial humic components are the most temperature dependent fractions during the BF processes as their sorption characteristics are negatively correlated with temperature. In general, it can be concluded that BF is capable to remove labile compounds under oxic conditions at all water temperatures; however, its efficiency is lower for humic compounds at higher temperatures.

2.1 INTRODUCTION

Pollution of surface water sources and the high cost of treatment have obliged water authorities to extend the use of cost-effective treatment techniques. Therefore, bank filtration (BF) has gained widespread interest in recent years as an economic surrogate for traditional drinking water treatment (Stahlschmidt et al., 2015). This technique has been employed in many European countries as a common method to supply drinking water. Many cities around the Rhine, Elbe, and Danube Rivers were primarily supplied with bank filtrate water for hundreds of years (Hiscock et al., 2002; Tufenkji et al., 2002). In recent years, BF has been utilized to contribute to the overall drinking water production in many developing countries: e.g., Egypt (Bartak et al., 2014), and India (Boving et al., 2014) with variable hydrological and environmental conditions. Thus, there is a need to evaluate the effectiveness of the BF process under these hot arid climates conditions. BF is a natural water treatment system in which surface water is induced to flow through a porous media towards a vertical or horizontal pumped well in response to a hydraulic gradient (Hiscock et al., 2002). The riverbed and the underlying aquifer had been proven to act as a natural filter to remove chemical and biological pollutants from the surface water system and thereby improve the pumped water quality. Moreover, the biochemical and physical processes (i.e., adsorption) that occur during subsurface flow have a substantial role in pollutant attenuation (Ray et al., 2002). The biochemical process taking place during infiltration is mainly controlled by the abundance and composition of dissolved organic matter (DOM) during the filtration process.

Natural water bodies contain a multitude of DOM types which determine the efficacy of the treatment processes in engineered and natural treatment systems (Baghoth et al., 2011). The organic matter present in surface water systems can be divided into two main categories: (i) non-biodegradable matter (e.g., humic substances HS) which is mainly formed from the decay of animals and plants in the environment, and (ii) biodegradable matter (e.g., protein-like compounds) which principally discharges into the water system from wastewater treatment plants (Nam et al., 2008). Although DOM doesn't have an adverse effect on human health, it impacts negatively the physical properties of the water (e.g., odour, taste and colour). In addition, it is considered the precursor for disinfection by-products (DBPs) carcinogenic compounds formation (Zhang et al., 2016). Furthermore, DOM components play major roles in the removal of pollutants during the treatment processes (Baghoth et al., 2011). Ma et al. (2018) reported that HS play influential role in the biodegradation of organic micropollutants (e.g., estrogen) in the treatment systems. Due to its high shuttle-electrons capacity, HS might enhance the bacterial growth and thereby the biotransformation of these micropollutants in treatment systems. Moreover, it can act as a redox mediator and thereby stimulating the iron and manganese microbial reduction process and enhancing the release of toxic metals (e.g., As and Cd) from sediment into the filtrate water in natural treatment systems (Vega et al.,

2017). Recently, Chianese et al. (2017) stated that HS absorbs a wide range of wavelengths of UV radiation and thus reducing the available energy for photo-degradation of organic micropollutants. Biodegradable matter, on the other hand, takes part in the following processes: (i) it enhances biofouling in reverse osmosis, nanofiltration and ultrafiltration membranes (Shi et al., 2018), (ii) it is used as a substrate for microorganism regrowth in distribution systems, (iii) it serves as precursor to nitrogenous DBPs (N-DBPs) formation in conventional treatment plants (Rostad et al., 2000).

BF was reportedly effective at reducing the labile organic compounds during infiltration and thus increasing the biological stability of drinking water in distribution systems by >60% as well as reducing the potential for DBPs formation by 40-80% (Drewes et al., 2003). The natural attenuation of DOM during BF is primarily due to initial adsorption followed by biodegradation (Gross-Wittke et al., 2010). These processes are highly influenced by subsurface flow area environmental conditions (i.e., temperature, redox conditions, travel time, raw water quality) (Pan et al., 2018). Maeng et al. (2008) found that more than 50% of the DOM is principally removed during the first 50 cm of infiltration and thus it is highly controlled by raw water temperature. Temperature may affect the DOM behaviour directly by altering the associated soil microbial activity and changing the pollutant adsorption character. Indirectly, DOM may reduce the dissolved oxygen in the infiltrate water and thus increase the potential for developing anoxic and even anaerobic environments in the adjacent aquifer. Adversely, redox alteration may impact the DOM biodegradation rate (Diem et al., 2013). Hoehn et al. (2011) reported the redox environment turning to Mn(III/IV)- and Fe(III)-reducing conditions during the hot summer of 2003 along the Thur River. Derx et al. (2012) observed that rising water temperature leads to lower water viscosity thereby increasing infiltration capacity and shortening travel time, which inversely affects the chemical pollutant removal efficiency. Ray et al. (2002) reported that the impact of temperature on water viscosity doubled the infiltration capacity during summer along the Ohio and Danube Rivers. However, this research focussed on the direct influence of temperature and redox conditions on DOM removal during BF processes.

Several field and lab-scale studies have tracked the behaviour of DOM during BF processes (Derx et al., 2012; Diem et al., 2013; Maeng et al., 2010). However, most research was conducted under cold and moderate-temperature (5-25°C) conditions. The bank filtrate temperature was recently recorded as 26.4°C along the Nile River in Egypt (Ghodeif et al., 2016) and 30°C along the Yamuna River in India (Sprenger et al., 2012). Moreover, recent climate models predict an increase in average global temperature by 1.4–5.8 °C by 2099 (Misra, 2014). Therefore, it is important to assess the effectiveness of BF to remove DOM under these extreme hot climate conditions. The main objectives of this research were: (1) to study the impact of high temperature (20-30°C) on bulk organic matter removal during BF processes; (2) to track the behaviour of the DOM fractions during BF processes using innovative analytical tools (i.e., fluorescence spectroscopy coupled with parallel factor analysis (PARAFAC) and liquid chromatography with an on-line organic carbon detection (LC-OCD/OND); (3) to determine which DOM fraction is more impacted by the temperature change and redox conditions; and (4) to quantify the role of biodegradation in DOM removal. To achieve these objectives, laboratory-scale batch studies were conducted to assess the impact of temperature (20, 25 and 30°C) on DOM behaviour using different influent water sources. Additionally, the impact of redox conditions on the reduction of DOM during BF was tracked in laboratory-scale soil columns at controlled room temperature (30 ±2°C).

2.2 MATERIALS AND METHODS

2.2.1 Batch Experiments

Batch experiments were conducted to study the impact of temperature on effluent and DOM behaviour in a saturated subsurface flow system. The batch reactors were operated (in duplicate) using 0.5 L glass bottles filled with 100 g of sand (grain size 0.8-1.25 mm) and fed with 400 mL of Delft canal water. The characteristics of the sand is presented in Table 2.1. The reactors were placed on a horizontal reciprocal shaker (shaking speed 100 rpm). Three sets of batch reactors were used at three different temperatures (20, 25, and 30) ±2°C. Initially, the reactors were acclimated (with respect to DOC removal) at their respective temperature for 90 days. After the acclimation period, the reactors were fed with four different types of water having different organic matter composition: (1) Delft canal water, the Netherlands (DC), (2) Delft canal water spiked with secondary treated wastewater effluent from Hoek van Holland, The Netherlands (DCWW), (3) secondary treated wastewater effluent (WW) and (4) water extractable organic matter (WEOM). WEOM was used to simulate the DOM water with high concentration of humic aromatic compounds. It was prepared using 100 g of clay (obtained from Delftse Hout, Delft, Netherlands) in a 0.5 L glass bottle filled with 400 mL of DCW water and placed on a shaker at 150 rpm for 24 hours. Then, the extracted solution was centrifuged at 4800 rpm for 30 minutes, and filtered with 0.45-μm pore-size cellulose acetate filters (Guigue et al., 2014). Influent and effluent samples were taken and analysed to determine their chemical and physical characteristics. Control samples were taken by filling the glass bottles with the same amount of each influent (without silica sand). Another series of batch reactor studies were performed to estimate the role of biodegradation in the removal of organic matter and to what extent it may be affected by temperature. Maeng et al. (2011a) suggested sodium azide as a biocide to suppress biological activity. However, this research found that sodium azide enhances fluorescence intensity and UV-absorbance measurements thus reducing their reliability, as also reported by Park et al. (2018). Alternatively, the batch reactors were spiked with mercuric chloride $HgCl_2$ (20 mM) to develop an abiotic environment inside the reactors (Choudhury et al., 2018).

Table 2.1. Characteristics of the silica sand media used in the laboratory experiments

Parameters	Values
Bulk density (g/cm³)	1.50±0.12
Particle density (g/cm³)	2.57±0.08
Porosity	0.42±0.06
TOC (µg/g)	12.70±1.10
Zn(II) (µg/g)	54.00±1.01
Se(IV) (µg/g)	2.30±1.20
Cu(II) (µg/g)	74.00±0.70
Pb(II) (µg/g)	5.00±0.50
Ni(II) (µg/g)	5.00±0.70

2.2.2 Column Experiments

Laboratory-scale column study (Figure 2.1) was conducted to assess the impact of redox conditions on the removal of DOM at high temperature (30°C) during the BF process. Six columns were established and run under three different redox conditions (oxic, anoxic, and anaerobic). Each column was made of PVC pipe with a 0.05 m internal diameter and 0.5 m height. The column bottom was packed with a support layer of graded gravel (7 cm high), and then with cleaned silica sand (size 0.8-1.25 mm, 40 cm height) allowing the media to settle in deionized water ensuring packing homogeneity. The columns were operated in up-flow mode (saturated flow), where a variable speed peristaltic pump was connected to the bottom of each column to introduce the influent water from the tank into the column at a constant hydraulic loading rate of 0.5 m/day. Two valves were attached

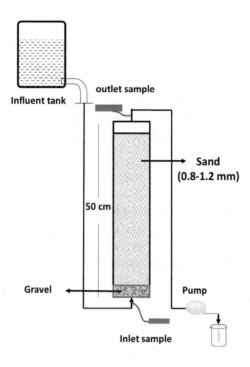

Figure 2.1. Schematic diagram of the soil column experimental setup

at the inlet and outlet of each column that allowed the air to dissipate from the system as well as to collect samples of the influent and effluent water. The oxic environment was maintained through aeration of the influent tanks continuously to keep the dissolved oxygen level at 7 mg/L. Anaerobic conditions were developed in the second two columns by degassing the influent tanks with nitrogen to dissipate the air. Anoxic conditions were created through the degassing processes followed by spiking 5 mg/L of nitrate into the influent tank. The columns were acclimated for 70 days until the removal of DOC for three successive measurements was ±1%. Then, three columns were fed with DCW and run under the identified redox conditions. The other three columns were fed with WEOM and run under the same redox conditions. All influents were filtered through a microsieve (38 μm) to avoid physically clogging the column inlets. The experiment lasted 30 days, and influent and effluent samples were taken daily.

2.2.3 Analytical methods

The collected samples were filtered using 0.45 µm filtration (Whatman, Dassel, Germany) and analysed within 3 days to avoid organic matter degradation Ammonium, nitrate and phosphate concentrations (mg/L) of the feed water was determined using ion chromatography (881 Compact IC pro, Metrohm anions, Swiss). Fe, Mn and Zn concentrations were measured using Inductively Coupled Plasma, Mass Spectroscopy (ICP-MS) (Thermo Scientific, XSeries II, Bermen, Germany).

DOC (in mg C L^{-1}) was measured through the combustion technique using a total organic carbon analyser (TOC-VCPN (TN), Shimadzu, Japan). UV-Absorbance at 254 nm UV254 [cm^{-1}] was measured using a UV/Vis spectrophotometer (UV-2501PC Shimadzu). Specific ultraviolet absorbance SUVA254 (L/mg-m) was used as an indicator for the aromaticity degree and unsaturated structures of the bulk organic matter. It was determined by dividing the UV254 by its corresponding DOC measurement. Adenosine triphosphate (ATP) was measured as an indicator for microbial activity associated with the sand. The sampling and preparation protocols of ATP measurements are explained in Maeng et al. (2008). Details of the ATP extraction procedures and detection method employed is described in Abushaban et al. (2017).

The constituents of bulk organic matter were elucidated using different analytical methods including: Liquid chromatography–organic carbon and nitrogen detector (LC-OCD) (DOC-LABOR Dr. Huber, Karlsruhe, Germany) and fluorescence excitation-emission spectrophotometer (EEM). LC-OCD is an analytical chromatographic technique coupling with three types of detectors: organic carbon detector (OCD), ultraviolet absorbance detector (UVD) and organic nitrogen detector (OND) to separate the pool of DOC into 5 major fractions: biopolymers BP, humic substances (humic and building blocks) HS, LMW acids (LMWa), LMW neutrals (LMWn) and hydrophobic organic carbon (HOC) based on their molecular weight distribution. The measurements procedures are described in detail by Huber et al. (2011).

The Fluorescence Emission Excitation Matrices (EEM) technique was widely used to characterize the bulk organic matter into three main components (humic-, fulvic- and protein-like fractions) (Baghoth et al., 2011). EEM measurements were conducted at excitation wavelengths from 240 to 452 nm with 4 nm intervals and emission wavelengths ranging between 290 and 500 nm with 2 nm intervals using a Fluoromax-3 spectrofluorometer (HORIBA Jobin Yvon, Edison, NJ, USA). The EEM spectrums were corrected as described in Abel et al. (2013). Briefly, Milli-Q water was used as a blank and subtracted from the EEM spectrums, the inner effect was minimized by dilution and the data were Raman normalized through dividing by the integrated area under the Raman scatter peak of Milli-Q water. The EEMs were corrected and recorded in Raman units (RU) using MATLAB (version 8.3, R2014a).

2.2.4 PARAFAC modelling

Fluorescence excitation-emission matrix spectroscopy coupled with parallel factor analysis (PARAFAC-EEMs) is used to decompose the EEMs to independent fluorescent components representing different NOM compositions. PARAFAC-EEMs have been extensively developed to characterize DOM behaviour in natural and treatment systems (Baghoth et al., 2011). PARAFAC is based on decomposing the fluorescence signals into tri-linear components and a residual array using an alternating least squares algorithm (Andersen et al., 2003):

$$X_{ijk} = \sum_{f=1}^{f} a_{if} b_{jf} c_{kf} + \varepsilon_{ijk}, \quad i = 1,\ldots\ldots,I; \; j = 1,\ldots\ldots,J; \; k = 1,\ldots\ldots,k; \; f = 1,\ldots\ldots,F$$

Where X_{ijk} represents the fluorescence intensity of the ith sample at the kth excitation and j^{th} emission wavelength; f is the number of model components; a_{if} is the score for the fth component and it is proportional to the fluorophore f concentration in sample i; b_{jf} is the scaled estimates of the emission spectrum for the fth component; c_{kf} is linearly related to the specific absorption coefficient at excitation wavelength k^{th}; ε_{ijk} is the residual term representing the unaccounted variation of the model (Stedmon et al., 2003).

To further assess the behaviour of different DOM components during the filtration process, a PARAFAC model was developed and validated using the complete measured F-EEMs dataset (184 samples) from the influent and effluent water of the batch and column experiments models following the steps proposed by Murphy et al. (2013). Briefly, an initial exploratory examination was first implemented to identify the poor quality measured data (outliers) and removed them from the dataset. The outliers (samples or variables) are commonly produced due to sampling or measuring errors and are determined by conducting the leverage analysis. The leverage value ranges between zero and one and expresses the deviation of measurement from the average data distribution. Then, models with different components (3-7) proceeded. The right number of PARAFAC components was selected and validated using diagnostic tools such as split-half validation (Colin et al., 2008), Tucker's congruence coefficients (Jason et al., 2010), and the residual error technique (Cuss et al., 2016). Split-half validation technique was used to validate the fluorescence model. This technique is based on the comparison of multiple models generated by splitting the dataset. In this research, the dataset was divided into four groups, and the samples were assigned alternately into the groups. Then, the groups were assembled into six combined groups; each combined group contains half of the dataset. After that, the PARAFAC test was applied for each combined group and the produced models were compared (Harshman et al., 1994). The PARAFAC model was implemented using the N-Way and drEEM MATLAB toolboxes developed by Murphy et al. (2013).

2.2.5 Statistical analysis

A two-way ANOVA test was applied to assess if an environmental parameter's influence on the DOM constituent behaviour during the BF process was statistically significant, in which for a significant difference $p<0.05$.

2.3 RESULTS

2.3.1 PARAFAC components

PARAFAC analysis successfully decomposed the fluorescence measurements into 5 components. The validated model explained more than 99.6% of the data variance. The excitation and emission loadings as well as the contours plots of these fluorescent components in RU are shown in Figure 2.2. The contour plots and corresponding excitation (solid curves) and emission (dotted curves) loadings of the fluorescent components C1-C5 identified from the complete F-EEMs dataset for the influent and effluent water of the batch and column studies.. The spectral slopes of the identified components were successfully cross-referenced with the OpenFluor database (Murphy et al., 2014) (Table 2.2).

Four of the PARAFAC components were identified previously as humics: (i) Component 1 (PC1) found at ($\lambda_{ex}\sim240$ and 320, $\lambda_{em}\sim410$ nm) and Component 2 (PC2) ($\lambda_{ex}\sim244$ and 376, $\lambda_{em}\sim480$ nm) are both associated with humic-like fluorophore substances originating from terrestrial resources as reported previously in Shutova et al. (2014). It can be seen that component 2 (PC2) appeared at longer excitation and emission wavelengths suggesting it possesses a more condensed and conjugated structure. According to Baghoth et al. (2011), these components are characterized by high molecular weight (>1000 Da). Moreover, they have low biodegradable matter and thus are principally removed by adsorption and coagulation in water treatment systems. (ii) Component 3 (PC3) ($\lambda_{ex}\sim300$, $\lambda_{em}\sim400$ nm) mimics microbial humic components in surface water systems This component is highly related to recent biologically produced fluorescent compounds (Walker et al., 2009). It is characterized by intermediate molecular weight (650<C3<1000 Da). (iii) Component 4 (PC4) ($\lambda_{ex}\sim360$, $\lambda_{em}\sim440$ nm) is related to a humic-like component derived from agricultural activity and it is common in freshwater environments as reported in Osburn et al. (2016b). These compounds mainly contain carboxylic and phenolic moieties in their structures (Singh et al., 2013). (iv) Component 5 (PC5) ($\lambda_{ex}\sim240$ and 270, $\lambda_{em}\sim320$ nm) is spectrally similar to a protein-like fluorophore (tyrosine and tryptophan compounds) identified in Kulkarni et al. (2017). These components are highly correlated with microbial activity in water systems and principally their removal in engineered water treatment systems is attributed to biodegradation (Baghoth et al., 2011). Therefore, it can be used as a surrogate for tracking the bioavailable DOM during filtration.

To further investigate the behaviour of the DOM fractions during the filtration process, the maximum fluorescence intensity (F_{max}) was used to characterize the influent and effluent water and to track the behaviour of PARAFAC components under different environmental conditions. F_{max} fluorescence intensities give an estimation of the proportional contribution of each component to the full fluorescence spectra. This contribution highly relies on the DOM source and the behaviour of the fluorescent components during the filtration process (Baghoth et al., 2011).

Table 2.2. The spectral slopes of the identified components and their corresponding components in previous studies from the OpenFluor database

	Ex. wavelength (nm)	Em. wavelength (nm)	Similarity score	previous study	Component type
PC1	248,324	412	0.96	(Graeber et al., 2012)	Terrestrial humic
PC2	244,376	482	0.99	(Shutova et al., 2014)	Terrestrial humic, Higher aromaticity
PC3	300	406	0.98	(Walker et al., 2009)	Microbial humic
PC4	356	436	0.97	(Osburn et al., 2016b)	Terrestrial humic, lower aromaticity
PC5	240,272	322	0.99	(Kothawala et al., 2014)	Protein-like

2.3.2 Impact of temperature and influent organic composition on DOM behaviour (Batch experiments)

During this research, laboratory-scale batch studies were employed to assess the impact of temperature (20, 25, 30±2°C) on the removal of DOM during the filtration process.

25

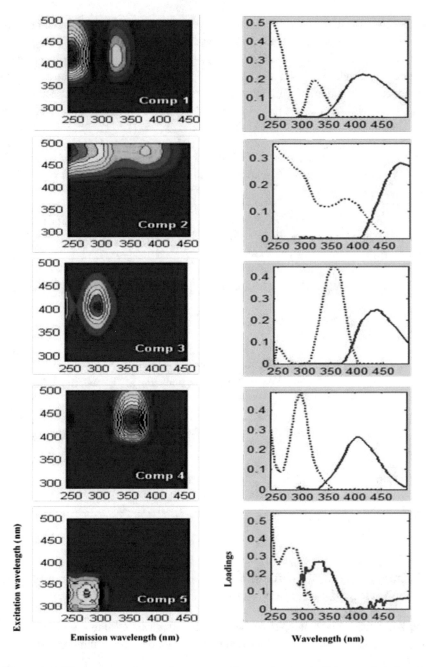

Figure 2.2. The contour plots and corresponding excitation (solid curves) and emission (dotted curves) loadings of the fluorescent components C1-C5 identified from the complete F-EEMs dataset for the influent and effluent water of the batch and column studies.

Characteristics of influent water DOM

The feed water quality has a clear impact on microbial activity and thus on DOM behaviour during the filtration process (Maeng et al., 2008). Four different water types were prepared and applied to the batch reactors. The average values of the chemical and physical water quality parameters are presented in Table 2.3. The results show that WEOM influent water had the highest DOC concentration (14.6 ±1.6 mg-C/L) followed by DCW (11.6±0.7 mg/L), DCWW (10.5±0.4 mg/L) and WW (9.7±0.6 mg/L). Furthermore, the WEOM had a relatively higher $SUVA_{254}$ value (3.56±0.71 L/mg-m) compared to DCW (2.84 ±0.33 L/mg-m). This implies that the WEOM influent was composed of higher amount of aromatic compounds (i.e., humic substances) than the DCW influent water. The average $SUVA_{254}$ values of DCWW and WW were 2.37±0.28 and 2.56±0.42 L/mg-m, respectively, implying relatively low aromatic character of their DOM composition.

Table 2.3. Chemical and physical characteristics of the influent water

	unit	DCW	DCWW	WW	WEOM
pH	---	7.87	7.79	7.66	7.65
DOC	mg-C/L	11.6±0.7	10.5±0.4	9.7±0.6	14.6±1.6
SUVA	L/mg-m	2.84±0.33	2.37±0.28	2.56±0.42	3.56±0.71
NO₃-N	mg-N/L	2.06±0.27	2.10±0.19	1.87±0.15	4.03±0.43
NH₄-N	mg-N/L	0.24±0.04	0.21±0.07	0.17±0.03	0.31±0.06
Mn	µg/L	46.8	14	14.03	86.74
Fe	µg/L	175	87.4	37.6	109.6
Zn	µg/L	20.9	30.1	36.6	36.6

LC-OCD/OND results showed that humic substances (HS) are the dominant fraction of DOC in all influent water. The contributions of HS to total DOC were 74%, 73%, 68% and 75%, respectively for DC, DCWW, WW and WEOM influent. The hydrophobic fraction (HOC) was only 5.7% of the DOC in WEOM influent water, a typical value for surface water systems. However, the HOC of DC, DCWW and WW influent was 10.3, 9.8, and 13.7% of the total DOC, respectively. This indicates the impact of EfOM on their organic compositions (Huber et al., 2011). Though WEOM had the highest concentration of BP, only 42% can be considered protein (assuming the C:N is 3, and all organic nitrogen in BP originates from protein) (Rehman et al., 2017). However, the protein represents 51, 65 and 82% of the BP for DC, DCWW and WW influent, respectively, which also reflects the impact of EfOM.

PARAFAC components (PC1-PC5) were recorded for all the influents. F_{max} was lower for protein component PC5 than the humic/fulvic components (PC1-PC4). The maximum and minimum F_{max} of component PC5 was observed for WEOM (1.09 ± 0.05 RU) and DCW (0.38 ± 0.03 RU), respectively. A humic-like component (PC4) exhibited a comparable contribution with a protein-like component to the DOM fluorescence of the influent. The F_{max} of PC4 ranged between 0.33 ± 0.04 and 0.89 ± 0.1 RU. However, the terrestrial humic-like component (PC1) contributed much higher than other humic/ fulvic components. An exception was the WEOM influent which possessed the highest concentration of conjugated humic component (PC2). Microbial humic (PC3) contributed moderately to the fluorescence spectrums of the feed waters, with higher contribution (1.39 ± 0.28 RU) observed for WW influent and lower contribution (0.99 ± 0.15 RU) for WEOM.

Bulk organic matter parameters

The results demonstrated that the DOC removal during the filtration process is highly dependent ($p<0.001$) on its concentration in the feed water. Table 2.4 showed that the DOC removal for DC, DCWW and WW influents were 9.5, 11.4 and 14.7%, respectively at 30°C. However, WEOM influent water exhibited the highest DOC removal (44%) at the same temperature and that may be attributed to higher feed water DOC concentration promoting biomass formation associated with sand. The ATP of reactor media were measured to be 4.69, 5.21, 5.39, and 7.95 µg/g sand at 30°C for DC, DCWW, WW and WEOM, respectively. These values increased by 7-9% at 25°C and 8-16% at 20°C (Figure 2.3). However, the statistical analysis revealed that there is no significant ($p>0.05$) effect of temperature on biological activity (ATP concentration) and thereby DOM biodegradation is not significantly affected by temperature in the range of 20-30°C. Nevertheless, the results showed higher DOC removal efficiency at lower temperature (20°C) ($p<0.05$). For instance, the DOC removal for DCW increased from $9.5 \pm 2.3\%$ at 30°C to $20.3 \pm 3.7\%$ at 20°C.

On the other hand, the results of abiotic batch reactors revealed that adsorption mechanisms contributed to the overall removal of DOC for DCW influent by $18 \pm 2.1\%$ at 30°C, $38.5 \pm 5.4\%$ at 25°C and $51 \pm 4.7\%$ at 20°C and for WEOM influent by $27 \pm 3.7\%$ at 30°C, $42 \pm 5.1\%$ at 25°C and $58 \pm 6.8\%$ at 20°C (Table 2.4).

In the same regard, $SUVA_{254}$ values exhibited a positive relationship with temperature, increasing from 2.84 ± 0.3 L/mg-m for the DCW influent to 3.71 ± 0.3, 3.59 ± 0.5, and 3.57 ± 0.2 L/mg-m for the effluent water at 30, 25, and 20°C, respectively (Table 2.5). This implies that aromatic compounds are favourably removed at lower temperatures, considering that there is no significant change in the removal of aliphatic compounds at respective temperatures.

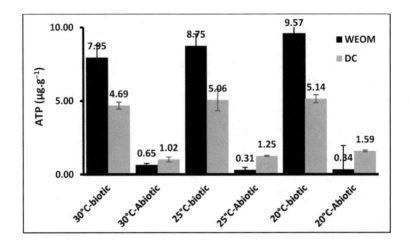

Figure 2.3. Changes of ATP concentrations (µg/g) in a function of temperature and biotic/abiotic condition during the batch experiment

Table 2.4. DOC (mg/L) values of the batch effluents at different temperatures (20, 25, 30°C) and biotic/abiotic oxic conditions

	30 °C		25 °C		20 °C	
	Biotic	**Abiotic**	**Biotic**	**Abiotic**	**Biotic**	**Abiotic**
DCW	10.5±0.21	11.4±0.35	9.76±0.0.28	10.89±0.61	9.24±0.32	10.4±0.43
DCWW	9.3±0.18	10.29±0.33	9.24±0.37	10.1±0.23	8.18±0.17	9.4±0.29
WW	8.27±0.24	9.42±0.51	7.95±0.27	9.37±0.23	7.37±0.19	8.96±0.18
WEOM	8.18±0.26	12.87±0.37	7.74±0.41	11.7±0.53	6.8±0.27	10.1±0.41

Table 2.5. SUVA (L.mg-1.m-1) values of the batch effluents at different temperatures (20, 25, 30°C) and biotic/abiotic oxic conditions

		DCW	DCWW	WW	WEOM
IC		2.84	2.37	2.56	3.56
30 °C	**biotic**	3.71	3.44	3.75	4.77
30 °C	**Abiotic**	3.68	3.69	3.72	3.83
25 °C	**biotic**	3.59	3.19	3.65	4.39
25°C	**Abiotic**	3.41	3.58	3.62	4.23
20 °C	**biotic**	3.57	3.18	3.59	4.26
20 °C	**Abiotic**	3.43	3.71	3.64	4.89

LC-OCD/OND analysis

Total DOC measured by LC-OCD/OND is largely well-matched with the measured values of a conventional TOC analyser to within ±0.5 mg/L. The DOC measured by LC-OCD/OND could be classified into: 1) chromatographic organic carbon (CDOC) that is calculated by area integration of the total chromatogram. 2) Hydrophobic organic carbon (HOC) that represents the difference between DOC and CDOC. It is anticipated that HOC remains in the columns due to its high hydrophobic interaction with the column media. A clear increase in HOC concentrations was observed, particularly at higher temperatures; HOC of influent DCW was increased by 64, 53, 23% at 30, 25, 20°C respectively. According to Tong et al. (2016), HOC is mainly composed of humins compounds which represents more than 50% of soil organic matter. Thus, it can be deduce that HOC increasing during the filtration process is mainly attributed to the leaching of organic matter from the soil into the effluent water, which is largely increased with temperature. Such increasing pattern was also observed in soil column experiments simulating BF by Filter et al. (2017). In contrast, CDOC (which represent the higher portion of influent DOC) was obviously decreased during the filtration process at all the respective temperatures. This removal is associated negatively with temperature; higher removals were obtained at 20 °C (50% for WEOM and 19% for DC) and lower removals were obtained at 32°C (44% for WEOM and 8% for DC). The removal of CDOC is highly dependent on the influent organic composition. The changes of CDOC fractions at different temperatures are presented in Figure 2.4 and Table 2.6. The removal of BP for DCW influent were (87, 94, 95%), DCWW (96, 91, 88%), WW (94, 86, 83%) and

WEOM (98, 97, 97%) at 30, 25, 20°C, respectively. However, the statistical analysis revealed that this process is independent of temperature (p>0.05).

Other biogenic organic matter fractions (LMWn and LMWa) exhibited lower removal for all the influent water compared to BP. The removal rates of LMWn for DC, DCWW, WW, and WEOM were 20, 16, 6, and 47% respectively at 30°C. This removal increased by 5-16% at 20°C. Likewise, the LMWa removal ranged between 10-47% for all the influents water at 20°C. This removal decreased by (6-25%) when the temperature went up by 10°C. This indicates that the decomposition of higher molecular weight compounds (BP and HS) into LMW hydrophilic compounds is lower than the removal of LWM compounds during filtration. However, this removal is also independent (p>0.05) of temperature. An exception was the LMW acid of WEOM, which increased by 24% at 30°C.

Humic compound removal exhibited a significant dependency on temperature (p<0.05). Figure 2.4 displays highly reduced HS concentration at lower temperatures (20, 25°C). The HS removal varied between 7-44% at 20°C and 4-41% at 25°C for all the influents water. However, an increase in the HS concentration was observed for DC, DCWW and WW at 30°C.

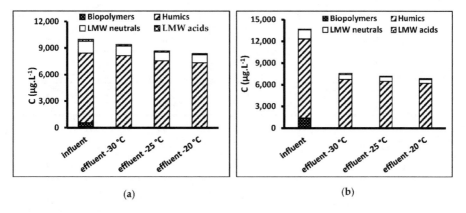

(a)   (b)

Figure 2.4. Changes in the LC-OCD/OND fractions concentration in batch reactors at different temperatures under oxic conditions: (a) DCW and (b) WEOM

Table 2.6. Changes in the LC-OCD fractions concentration in batch reactors at different temperatures under oxic conditions using different water types (DC, DCWW, WW and WEOM)

Feed water type		BP (μg.L^{-1})	HS (μg.L^{-1})	LMWn (μg.L^{-1})	LMWa (μg.L^{-1})
DCW	Influent	562	7860	1350	200
	Effluent-30 °C	74.7	8070	1080	172
	Effluent-25 °C	32.2	7520	993	145
	Effluent-20 °C	28.8	7310	907	123
DCWW	Influent	497	7670	1120	194
	Effluent-30 °C	20	7780	940	120
	Effluent-25 °C	45	7120	945	108
	Effluent-20 °C	61	5990	890	113
WW	Influent	360	6610	1200	224
	Effluent-30 °C	22	6710	1130	160
	Effluent-25 °C	51	5890	1060	133
	Effluent-20 °C	62	5540	940	118
WEOM	Influent	1420	10900	1250	118
	Effluent-30 °C	24.7	6730	665	146
	Effluent-25 °C	40.3	6440	614	91.8
	Effluent-20 °C	38.6	6160	542	106

PARAFAC- EEM analysis

The protein-like (PC5) component exhibited the highest reduction rate at all three temperatures (Figure 2.5). The protein-like component removal increased consistently with increasing influent concentration (p<0.001). The highest F_{max} reduction (93.6±2.6%) was recorded for WEOM at 20°C, followed by WW (60.2±3.2%), DCWW (43±3.8%) and DCW (36.1±1.7%), respectively, at the same temperature. Similar to BP removal, these labile compounds exhibited independent behaviour upon temperature variation

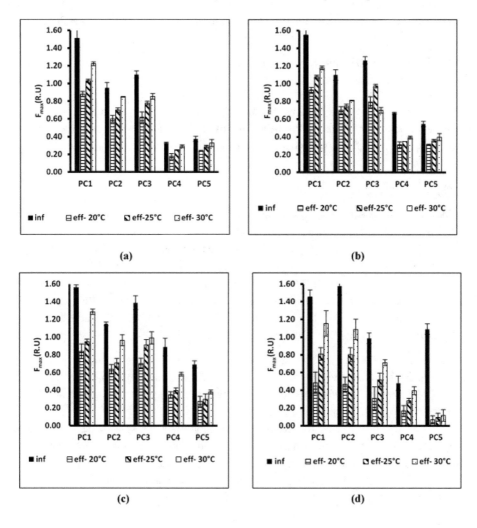

Figure 2.5. Changes of PARAFAC components (F_{max}) in batch reactors at different temperatures under oxic conditions: (a) DCW, (b) DCWW, (c) WW, (d) WEOM

(p>0.05). The removal percentage was reduced by 23.9±4.8%, 15.9±1.6%, 15.08±5.3%, and 4.5±1.87%, respectively, for DC, DCWW, WW, and WEOM at 30°C. On the contrary, Humic components (PC1-PC4) removals were impacted significantly by variations in temperature and feed water characteristics ($p<0.05$). An exception was microbial humic, which showed independent behaviour with temperature variations during the filtration process ($p=0.09$). Figure 2.5 illustrates the attenuation of humic components decreasing with rising temperature. The average removal of PC1 was 48.6±12.3, PC2 (47±16%), PC3 (49.8±13.5%) and PC4 (56.2±8.4%) at 20°C. These removals decreased by 28.4, 26, 19 and 30.4% for (PC1-PC4) respectively at 30°C.

2.3.3 Impact of redox conditions on DOM behaviour (Column experiments)

Column experiment was conducted to assess the impact of redox conditions (oxic, anoxic and anaerobic) conditions on the behaviour of DOM constituents during the filtration process. The experiment was conducted at a controlled temperature room (30±2°C) using two different feed water types (DCW and WEOM).

Characteristics of influent water DOM

The bulk organic characteristics of the feed water are presented in Table 2.7. It can be shown that WEOM had higher DOC concentration (14.16±0.73 mg/L) than DCW influent (10.80±0.51 mg/L). Moreover, WEOM possessed higher aromatic characteristics (SUVA = 3.67 ±0.21 L/mg-m) compared to DCW influent (SUVA = 3.05±0.31 L/mg-m). These results were confirmed with PARAFAC-EEM results which demonstrated that WEOM had higher concentration of terrestrial-derived; the average F_{max} of PC1, PC2 and PC4 for WEOM were 1.52±0.06, 1.61±0.1 and 0.51±0.03 RU for WEOM, and 1.31±0.04, 0.99±0.07 and 0.41±0.02 for DCW influent, respectively.

Furthermore, LC-OCD/OND results revealed that the humic fraction was the dominant fraction in the feed water, it represents 72 and 73% of the DOM pool for DCW and WEOM influents, respectively. However, biogenic fractions (BP, LMWn and LMWa) represent only 5.9, 11 and 2.3% of DOM for DCW influent and 10.7, 8.7 and 1.1% of DOM for WEOM influent, respectively. In addition, PARAFAC-EEM results revealed that DCW influent possessed higher concentration of microbial humic-like component (PC3) and lower of protein-like component (PC5); the F_{max} value of PC3 and PC5 were 0.98±0.07 and 0.58±0.03 RU for DCW and 0.78±0.06 and 0.92±0.04 RU for WEOM influent, respectively.

Table 2.7. Characteristics of the influents and effluents water of the columns under different redox conditions

	DC			WEOM		
	pH	DOC	SUVA	pH	DOC	$SUVA_{254}$
	----	(mg/L)	(L/mg-m)	----	(mg/L)	(L/mg-m)
Influent	7.82	10.80±0.51	3.05±0.31	7.73	14.16±0.73	3.67±0.21
effluent-oxic	7.91	9.49±0.36	4.01±0.24	7.88	7.91±0.17	3.92±0.14
effluent-anoxic	8.08	10.12±0.25	3.06±0.19	8.16	9.37±0.28	3.73±0.33
effluent-anaerobic	8.13	10.26±0.55	3.21±0.27	7.95	10.08±0.39	3.67±0.37

Bulk organic matter parameters

The redox environment significantly ($p<0.05$) impacts the removal of DOC during filtration. Table 2.7 showed that the removal of DOC decreased by 5-10% under anoxic, and 7-15% under anaerobic conditions compared to oxic conditions. This is highly linked

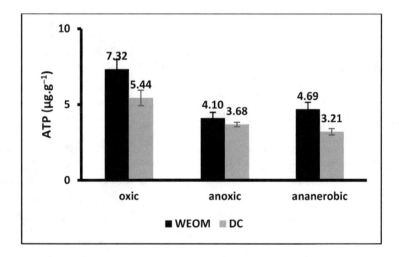

Figure 2.6. Changes of ATP concentrations (µg/g) in a function of redox conditions during the column experiment for DCW and WEOM influent

35

to the biological activity associated with the sand. ATP from active microbial biomass associated with sand was higher for oxic conditions. The average concentrations of ATP in the oxic, anoxic and anaerobic columns were 5.44±0.64, 3.68±0.37, and 3.21±0.47 for DCW and 7.32±0.51, 4.1±0.15, and 4.69±0.21 µg/g sand for WEOM, respectively (Figure 2.6). In the same regard, SUVA increased from 3.05±0.31 to 4.01±0.24, 3.06±0.19 and 3.22±0.27 L/mg-m for DC, and from 3.67±0.21 to 3.92±0.14, 3.73±0.33 and 3.67±0.37 L/mg-m for WEOM, respectively, under oxic, anoxic and anaerobic conditions.

LC-OCD/OND analysis

LC-OCD/OND results showed that BP is preferentially removed during soil passage; its removal under anoxic and anaerobic conditions was less than oxic conditions by 20-24% (Figure 2.7). Similarly, LMWn compounds exhibited higher removal during oxic filtration. The removal was decreased under anoxic and anaerobic condition by 21 and 50% for WEOM and by 15 and 17% for DCW, respectively. The same behaviour was observed for LMWa of DCW influent, where its removal decreased by (25-32%) under sub-oxic condition. However, LMWa of WEOM influent exhibited inconsistent behaviour, where its concentration was increased by (15-21%) under sub-oxic conditions compared to its concentration in the feed water. In the same way, HS demonstrated more removal efficiency under oxic condition. For WEOM, the average removal of HS under oxic, anoxic and anaerobic conditions was 36, 29, and 24%, respectively. However, HS

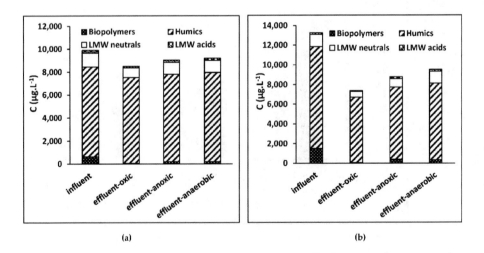

(a)

(b)

Figure 2.7. Changes of LC-OCD/OND fractions under different redox conditions: (a) DCW, (b) WEOM (temperature = 30°C, column study)

removal for DCW was decreased only by 2-4% when the environment turned into sub-oxic conditions.

PARAFAC-EEM analysis

The fate of PARAFAC components under different redox conditions was examined using their maximum fluorescence intensity (F_{max}). The results reveal that the redox environment plays a substantial role in the removal efficiency of fluorescence components during soil passage (Figure 2.8).

The removal of protein-like component was decreased by 15-22% under sub-oxic conditions. In the same manner, microbial humic component (PC3) displayed redox-dependent behaviour, with higher reduction under oxic condition. The removal of microbial humics were reduced by 20-22% for WEOM and 2-5% for DCW influent under sub-oxic conditions. Similar to protein-like component, terrestrial-derived humic components showed higher removal under oxic condition; the removals of PC1, PC2 and PC4 were reduced by 14-19%, 9-14% and 10-18%, respectively during sub-oxic filtration.

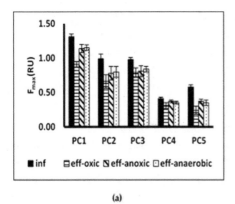
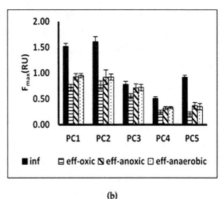

(a)   (b)

Figure 2.8. Changes of PARAFAC components (F_{max}) under different redox conditions: (a) DCW, (b) WEOM (temperature = 30°C) (column study)

2.4 DISCUSSION

2.4.1 Impact of temperature and influent organic composition on DOM behaviour

DOM removal during BF is principally due to a combination of DOM sorption to the sand media and biodegradation through bacteria in biofilms associated with the media. In this study, DOC decreased during the filtration process, showing a high dependence on the feed water DOC composition, which is highly correlated with the biomass activity associated with the sand. This is in agreement with Li et al. (2012) which reported a positive correlation between biofilm density and influent DOC concentration. However, the results refer that there is no effect of the temperature on the biomass activity associated with the sand and thus the impact of temperature on DOC biodegradation is low. On the other hand, the abiotic experimental results in this study indicated preferential DOM adsorption at lower temperature. Thus, it can be concluded that the relatively higher removal of DOM at lower temperatures in the range of 20-30°C during the BF process is mainly ascribed to adsorption. Massmann et al. (2006) reported that DOM attenuation is independent of temperature between 0-24°C, based on a field study conducted at an operational artificial recharge site over Tegel Lake in Germany. In contrast, Abel et al. (2012) found a positive relationship between DOC removal and temperature (5-25°C) during the filtration process. Alidina et al. (2015), on the other hand, reported a minor temperature (10-30°C) effect on DOM removal during a column filtration process, with higher removal at lower temperatures. These contradictions in DOM behaviour during infiltration at different temperatures could be attributed to different feed water DOM characteristics. This goes in line with the conclusion of Chen et al. (2016) who reported that the organic composition of the raw water determines its behaviour during filtration. Thus, it is important to assess the behaviour of DOM fractions individually during filtration.

BPs (MW > 20,000 Da) are the most readily biodegradable DOM fraction, thus they are preferentially removed during the filtration process (Vasyukova et al., 2014). According to So et al. (2017), the BP was reported to be highly degraded at higher temperatures, however, in this research, there is no significant effect of temperature (20-30 °C) on BP removal observed during filtration. The removal of BP reached >80% at all temperatures during filtration, this is mainly attributed to the prolonged filtration period. Moreover, the high ratio between organic nitrogen BP and DOC (OND/DOC) in the feed water indicated that they are principally composed of proteinaceous matter that is highly degraded during filtration. This in agreement with the PARAFAC-EEM results, where labile compounds (i.e., protein-like compounds) exhibited independent behaviour upon temperature variation. However, the results demonstrated that protein-like component removal is more sensitive to temperature variation when the protein content of the feed water is low. This specifies the vital role of co-metabolism in the removal of this biodegradable matter. These results are consistent with Maeng et al. (2010) who reported the total removal of BP during BF over Tegel Lake (Germany). Likewise, other biogenic organic matter fractions (LMWn and LMWa) exhibited temperature-independent behaviour during the filtration process.

In contrast, the removal of humic compounds was highly dependent on the temperature of the feed water, with a favourable reduction at lower temperature (20-25°C). This reduction is mainly ascribed to the high ability of these refractory compounds to adsorb onto sand grains at lower temperature. On the contrary, an increase in the concentration of HS was observed at higher temperature (30°C) for DC, DCWW and WW influent. The HS concentration enrichment may be attributed to microorganisms and enzymes that are able to: (i) transform microbial matter to more refractory and conjugated matter (i.e., microbial humification process), that was reported in several laboratory-scale and field studies (Jørgensen et al., 2011; Yang et al., 2013; Yang et al., 2014). (ii) leach soil humic compounds into filtrate water (Sun et al., 2017). In contrast, the humic of WEOM influent exhibited a unique behaviour, its concentration decreased at all three temperatures. This may be due to: (i) the absence of microorganisms to transform labile matter or leach organics from soil as mentioned above. (ii) The inability of these organic compounds to bio-transform into refractory compounds, may be attributed to higher aromaticity (SUVA = 2.9 L/mg-m) compared to other influent. PARAFAC-EEM results revealed that the condensed structure humic compounds are the most impacted by changing temperature during filtration process. These results are compatible with those of Abel et al. (2012), who illustrated that the optimum temperature for removal of these refractory compounds is 15°C. The ratios between the PARAFAC components were used in many studies (Baghoth et al., 2011; Kulkarni et al., 2017) to assess treatment efficacy. In this research, only the ratios between the F_{max} of terrestrial humic components (PC1, PC2, and PC4) and protein-like component (PC5) exhibited a clear increasing trend with rising temperature. These results also confirm the preferential removal of terrestrial humic components at lower temperatures.

2.4.2 Impact of redox conditions on DOM behaviour

This research illustrated the preferential removal of DOM under oxic condition during the BF process, which is mainly attributed to oxygen as an electron acceptor for microorganism respiration to degrade the organic matter. Slower biodegradation of DOM under sub-oxic conditions was also reported in previous studies (Abel et al., 2012; Gimbel et al., 2006). Moreover, SUVA values exhibited lower increase during sub-oxic condition, which refers to the preferential removal of aliphatic compounds during oxic filtration.

LC-OCD/OND data revealed that BP is the most impacted DOM fraction by the alteration in the redox environment, with favourable removal under oxic condition. In the same regard, the ratio between nitrogen and carbon BP exhibited higher values under oxic conditions, which infers lower biodegradation of protein compounds under sub-oxic conditions. This finding was confirmed with PARAFAC-EEM results, where F_{max} of the protein-like component (PC5) exhibited higher reduction (relative to influent F_{max}) under oxic conditions than other redox conditions. This reduction is mainly ascribed to the degradation of high molecular weight biodegradable organic matter into non-fluorescing material. This is in agreement with field data collected at the Tegel Lake (Berlin, Germany) BF site where partial removal of BP was detected under sub-oxic conditions (Gimbel et al., 2006). Furthermore, previous studies (Abel et al., 2012; Maeng et al., 2008) also emphasized the superior removal of protein-like components under oxic environmental conditions by conducting laboratory-scale experiments. Likewise, LMW (acids and neutrals) exhibited lower removal efficiency under sub-oxic conditions. An exception was the LMW acid of WEOM, which increased under sub-oxic conditions likely due to the breakdown of larger molecular weight humic matter into lower molecular weight compounds under these conditions (Wang, 2014).

HS removal followed the same trend as BP (with less removal efficiency), in that higher removal was obtained under oxic than other redox conditions. Nonetheless, HS removal efficiency is much less for DCW influent than WEOM influent, presumably attributed to the nature and molecular weight of the humic present. PARAFAC-EEM humic components (PC1-PC4) exhibited as well higher reduction under oxic conditions. Terrestrial humic-like components (PC1, PC2) exhibited the highest reduction of F_{max} among other humic components, followed by lower aromatic-humic such as component (PC4). According to Gerlach et al. (1999), humic compounds with higher molecular weight are preferentially removed during aerobic soil passage.

2.5 CONCLUSIONS

Based on the results of laboratory-scale batch and column studies, the following conclusions can be drawn:

- A positive correlation was found between DOM biodegradation and raw water DOM concentration, which was likely due to higher microbial activity associated with sand as determined by ATP measurements of the biomass attached to the sand grains.

- The removal of DOM during (bank) filtration is significantly impacted by temperature variation, with higher removal at lower temperatures.

- The experimental results revealed that the labile compounds (i.e., biopolymers) are highly removed (>80%) under oxic filtration regardless of the temperature and organic matter composition of the feed water.

- Humic substances removal exhibited a significant dependence on temperature, with higher removal at lower temperature. PARAFAC analysis indicated that terrestrial humic components are the least persistent humic type adsorbed at lower temperature. The contradictory behaviour of protein-like and humic compounds explains the positive relationship between SUVA and temperature.

- DOM was preferentially removed under oxic conditions; its removal decreased by 5-10% under anoxic, and by 7-15% under anaerobic conditions. LC-OCD/OND results reveal that biopolymers are the most impacted fraction by altering the redox conditions. Humic substances as well exhibited lower removal efficiency (with less extent) under sub-oxic conditions. Therefore, post-treatment steps should be considered in case of sub-oxic filtration.

- In general, this study revealed that the DOM components removal efficiency during BF under hot arid conditions (high temperature) is determined by the feed water DOM composition and redox conditions in the infiltration area.

Finally, this study shows that PARAFAC-EEM and LC-OCD/OND can be promising tools to provide further insight into BF processes and for determining the treatment efficiency for DOM components.

3

REMOVAL OF ORGANICS MICRO-POLLUTANTS DURING BANK FILTRATION

ABSTRACT

Riverbank filtration (RBF) represents a low-cost and sustainable alternative to advanced treatment technologies to pre-treat or remove several organic micropollutants (OMPs) from surface water. The objective of this research was to investigate the efficacy of biodegradation and adsorption processes in the removal of OMPs at high temperatures (20-30±2°C) during RBF. Laboratory-scale batch studies were conducted using silica sand at different temperatures (20, 25 and 30°C) to study the removal of 19 OMPs (6 polyaromatic hydrocarbons (PAHs), 8 herbicides and 5 insecticides) from various water sources with different organic matter characteristics. Simazine, atrazine, metolachlor, and isoproturon exhibited partial persistent characters (16% < removal < 59%), which apparently decreased with increase in temperature. DDT, pyriproxyfen, pendimethalin, β-BHC, endosulfan sulfate and PAHs with high hydrophobicity (solubility in terms of logS < -4) tend to be well adsorbed onto sand grains (removal > 80%), regardless of temperature, redox conditions or type of organic carbon fraction fed to the batch reactors. These findings indicate that these hydrophobic compounds are effectively removed during RBF regardless of the environmental conditions. Hydrophilic compounds (molinate, dimethoate, and propanil) showed temperature-dependent characteristics for influent water with low organic matter; their attenuation increased at higher temperature (removal > 95%) due to the high microbial activity. This study revealed that temperature is an important parameter affecting the removal of OMPs with hydrophilic and low-hydrophobicity characters. However, temperature has less influence on the removal of highly hydrophobic OMPs during RBF and thus should be considered during RBF system design.

3.1 INTRODUCTION

The need for high-quality drinking water is increasing rapidly worldwide as a result of increasing urbanization and population growth. However, contamination of surface water resources by domestic and industrial wastewater increase the costs of treatment. Most developing countries employ conventional water treatment techniques, including coagulation, flocculation, clarification, and filtration processes, and these methods are usually followed by disinfection with chlorine. However, conventional water treatment is not effective enough in removing persistent compounds, such as organic micro-pollutants (OMPs) (Maeng et al., 2013). Bank filtration (BF) could be a viable and cost-effective treatment option schemes to reduce the concentration of these persistent pollutants, as well as removing pathogens, algal toxins, and organic matter (Hamann et al., 2016; Hiscock et al., 2002).

OMPs removal during BF is mainly attributed to adsorption and biodegradation, which are highly influenced by the surrounding environmental conditions including temperature, redox and organic composition of the source water (Bertelkamp et al., 2014). Various studies were conducted using laboratory-scale experiments to investigate the effect of organic matter composition of influent water on the removal of OMPs during BF. According to Lim et al. (2008) and Maeng (2010), there is a positive relationship between biodegradable dissolved organic carbon (BDOC) concentration in the feed water and OMP removal, while other studies expound a negative relationship (Bertelkamp et al., 2016a; Li et al., 2014). This contradiction can be explained in terms of different OMP classes and characteristics, influent water characteristics and biological activity that is largely affected by the surrounding climate and environmental conditions. In the same regard, temperature may also directly affect the performance of BF by altering the soil sorption characteristics and exert indirect effects through alternating the redox conditions in the infiltration area, thereby impacting the removal of OMPs (Rohr, 2014). During BF, the initial infiltration zone is always characterized by oxic conditions followed by anoxic (NO_3^- as electron acceptor) and anaerobic conditions (Fe^{3+}/Mn^{4+} as electron acceptor). The redox zone size is largely based on the aquifer characteristics and the raw water quality. Few studies have been developed to determine the effect of redox conditions on OMP removal during BF processes. Maeng (2010) illustrated that 17α-ethinylestradiol (EE2) and 17β-estradiol removals were not significantly affected by the redox conditions. Bertelkamp et al. (2016a) reported that some OMPs including dimethoate, diuron, and metoprolol showed redox-dependent removal behaviour in favour of oxic conditions. However, these studies were undertaken at room temperature. Thus, it is important to assess the effect of redox conditions on the removal mechanisms of OMPs at high temperature.

During this research, laboratory-scale batch studies were conducted to understand the behaviour of different OMP classes (PAHs and pesticides) during BF and to assess the influence of different environmental conditions (temperature, redox, and influent organic matter composition) on their removals as well as to propose guidelines for OMPs removal prediction during BF processes at arid climate conditions.

3.2 MATERIALS AND METHODS

3.2.1 Experimental set-up

To assess the effect of temperature, redox conditions and organic matter composition on the removal of different classes of OMPs, 34 batch reactors were set up. 100 g of silica sand (grain size, 0.8-1.25 mm) were placed in 0.5-L glass reactors bottles. Initially, sand in the reactors was bio-acclimated with 400 mL of Nile River water mixed with secondary treated wastewater (1:1) under their respective environmental conditions of temperature (20, 25 and 30±2°C) and redox conditions (oxic, anoxic and anaerobic) (Table 3.1).The temperature levels were set based on field feedback from previous studies (Ghodeif et al., 2016; Sprenger et al., 2012). The acclimation period lasted for 60 days, the feedwater was renewed every 5 days until the reactors biologically stabilized (i.e., acclimated) with respect to dissolved organic carbon (DOC) removal.

3.2.2 Effect of organic composition and temperature on OMPs removal

After the acclimation period, the batch reactors were fed with different types of water of different organic matter composition (humics "humic and fulvic", and protein): Water from the Nile River at Gabal Taqoq region (Aswan, Egypt) (NR) was used to simulate influent water with low organic content. Other water types were developed by mixing the Nile River water with: i) secondary treated wastewater from Kima wastewater treatment plant at Aswan, Egypt (ST), ii) treated wastewater from oxidation ponds plant at Balana, Egypt (OP), and c) water extractable soil organic matter (WEOM). WEOM was prepared following the method of Guigue et al. (2014). The main purpose of using this type of influent water was to simulate the effect of influent water with a high humic concentration on OMP removal efficiency. All influents were filtered through a micro-sieve (38 μm) before application to the batch reactors. Each reactor was injected with the selected OMPs at a concentration of 5 μg/L each, this concentration is corresponded to the OMPs concentrations detected in the Egyptian surface water systems (Abdel-Halim et al., 2006; Dahshan et al., 2016; Mansour et al., 2003; Selim et al., 2009) and other surface water systems worldwide (Fadaei et al., 2012; Meffe et al., 2014). Then, the batch reactors were rotated on a shaker at 100 rpm and incubated at the desired temperature for 30 days. A blank experiment was undertaken by injecting the same concentration of the selected

OMPs in Milli-Q water, without sand and incubated under similar conditions, to investigate the loss of OMPs.

Table 3.1. Process conditions of batch experiments

Influents	Temperature	Redox	Biotic/abiotic
Nile River (NR)	20,25,30 °C	Oxic	biotic
Nile River: Secondary treated wastewater (ST)	20,25,30 °C	Oxic	biotic
Nile River: oxidation ponds treated wastewater (OP)	20,25,30 °C	Oxic	biotic
Nile River: Water extractable organic matter (WEOM)	20,25,30 °C	Oxic	biotic
Nile River (NR)	20,25,30 °C	Oxic	abiotic
Nile River (NR)	25 °C	anoxic/ anaerobic	biotic

3.2.3 Abiotic experiments

To assess the role of adsorption in the OMP removal process, 6 abiotic batch reactors were employed using NR influent water injected with 20 mM $HgCl_2$ to suppress biological activity (Choudhury et al., 2018). The experiment was conducted at three different temperatures (20, 25, 30±2°C).

3.2.4 Redox experiments

To investigate the effect of different redox conditions on the OMPs removal efficiency, 2 reactors were developed under anaerobic condition by degassing with a nitrogen stream to dissipate dissolved oxygen (DO) from the reactors (i.e., DO < 0.2 mg/L) (Maeng et al., 2011b), while another 2 reactors were run under anoxic conditions (nitrate-reducing environment) by degassing with a nitrogen stream and injecting 10 mg/L NO_3^-.

3.2.5 Organic micropollutants (OMPs)

The selected OMPs consisted of 6 polyaromatic hydrocarbons, 8 herbicides and 5 insecticides covering a wide range of physico-chemical properties (Table 3.2). All OMPs used were of analytical grade standard solutions and purchased from Accustandard, USA. Pesticides were analysed with LC-MS-MS following EPA Method 536. PAHs were measured using a GC-MS-MS instrument and following the EPA-625 method.

Table 3.2. List of OMPs studied and their properties

	Chemical structure	pKa Charge at pH 8	Water solubility (mg/mL)	LogS at pH 8	LogD at pH 8	Log P
Molinate	$C_9H_{17}NOS$	n.a	9.70E-01	-2.1	2.34	2.34
Simazine	$C_7H_{12}ClN_5$	3.23	6.40E-03	-3.45	1.78	1.78
Isoproturon	$C_{12}H_{18}N_2O$	n.a	7.20E-02	-3.08	2.57	2.57
Atrazine	$C_8H_{14}ClN_5$	3.2	3.47E-02	-3.8	2.2	2.2
Propanil	$C_9H_9Cl_2NO$	1.21	2.25E-01	0	0.56	3.85
Dimethoate	$C_5H_{12}NO_3PS_2$	-4.5	2.50E+01	-2	0.34	0.34
Pendimethalin	$C_{13}H_{19}N_3O_4$	-1	3.00E-04	-4.32	4.82	4.82
Metolachlor	$C_{15}H_{22}ClNO_2$	n.a	5.30E-01	-3.68	3.45	3.45
Pyriproxyfen	$C_{20}H_{19}NO_3$	2.86	1.63E-02	-5.11	4.75	4.75
Picloram	$C_6H_3Cl_3N_2O_2$	-0.21	4.30E+02	0	-1.43	2
DDT	$C_{14}H_9Cl_5$	n.a	8.50E-05	-6.55	6.46	6.46
Endosulfan sulfate	$C_9H_6Cl_6O_4S$	n.a	4.80E-04	-4.9	n.a	3.16
β-BHC	$C_6H_6Cl_6$	n.a	2.40E-04	-4.7	4.35	4.35
Naphthalene	$C_{10}H_8$	n.a	3.10E-02	-3.3	2.96	2.96
Fluorene	$C_{13}H_{10}$	n.a	1.69E-03	-4.8	3.74	3.74
Anthracene	$C_{14}H_{10}$	n.a	4.34E-05	-5.6	3.95	3.95
Pyrene	$C_{16}H_{10}$	n.a	1.35E-04	-6.9	4.28	4.28
Chrysene	$C_{18}H_{12}$	n.a	2.00E-06	-7.7	4.94	4.94
Benzo (b) fluoranthene	$C_{20}H_{12}$	n.a	1.50E-06	-7.7	5.27	5.27

3.2.6 Characterization of influents and effluents

The collected samples were stored at 4°C after 0.45-μm filtration (Whatman, Dassel, Germany) and analysed within 3 days of collection to minimize biodegradation of organic matter. Dissolved organic matter (DOC in mg C L^{-1}) was measured using a total organic carbon analyser (TOC-VCPN (TN), Shimadzu, Japan). UV absorbance [cm^{-1}] was monitored at 254 nm by a UV/Vis spectrophotometer (UV-2501PC Shimadzu). Specific ultraviolet absorbance ($SUVA_{254}$) [L mg^{-1} m^{-1}]), an indicator for the relative aromaticity and humic content of the bulk organic matter, was determined after dividing UV_{254} absorbance by DOC concentration. Characterization of organic matter fractions (humic, fulvic, and protein) was performed using a fluorescence excitation-emission spectrophotometer (F-EEM) as described in Baker (2001) study.

Peak picking technique was used to characterize the fluorescence excitation and emission matrices data of the influents and effluents water (Cheng et al., 2018; Coble et al., 2014). Three regions of peaks were identified at distinct excitation and emission wavelengths: primary humic-like OM peak (P1) was assigned at excitation=250 nm and emission=440 nm; (2) secondary humic-like OM peak (P2) was measured at excitation=320 nm and emission=440 nm; (3) protein-like OM peak (P3) was determined at excitation=240 nm and emission =330 nm. Furthermore, two fluorescence indices were used to characterize the organic composition of the feed water. (1) Humification index has used an indicator for the humic content of the feed water, it is calculated by dividing the peak area under the emission spectra 435–480 nm by the peak area 300–345 nm 435–480 nm, at excitation 254 nm (Coble et al., 2014; Ohno, 2002). (2) Fluorescence index (FIX) has used to identify the relative contribution of terrestrial (low values) and microbial (high values) dissolved organic matter to the full fluorescence spectrum. It is estimated as a ratio of the fluorescence intensity at emission wavelength 450 nm to that at 500 nm for a definite excitation wavelength (370 nm) (Coble et al., 2014; McKnight et al., 2001). FIX typically ranges between 1.2-1.7 for surface water systems, where high value is indicative for higher biodegradable content (Hansen et al., 2016). Nitrate and phosphate concentrations (mg/L) of the influent and effluent samples were quantified using ion chromatography (881 Compact IC pro, Metrohm anions, Swiss), whereas ammonium (mg/L) was determined using an ion chromatograph (881 Compact IC pro, Metrohm cations, Swiss). Heterotrophic plate counts (HPC) were used as an indicator of sand biomass activity following the method described in (Maeng et al., 2008). At the end of the experiment, duplicate wet sand samples (2–2.5 g) were suspended in 50 mL of autoclaved tap water and sonicated at a power of 40 W for 2 minutes to suspend the biomass into the solution. 0.1 mL of microbial suspensions were spread in triplicate over a surface of R2A agar plates and incubated for 48 hours at 35 °C.

3.2.7 Data analysis

Two-way ANOVA was used to determine the significance of the impact of temperature, influent organic matter and redox conditions on the removal of OMPs in batch reactors. The criterion level of significance (p) was 0.05.

3.3 RESULTS AND DISCUSSION

3.3.1 Characterization of influents and effluents water

Relevant water quality parameters of the feed water added to batch reactors and the average changes in their quality characteristics during the incubation period at different temperatures are summarized in Table 3.3. At all temperature set-points, OP water exhibited the highest DOC removal (51-54%) followed by ST water (46-48%), WEOM water (28-33%), and NR water (9-18%). The higher DOC reduction for OP and ST waters are mainly attributed to their high biodegradable organic matter concentrations, which support the biological activity associated with the sand.

The average concentrations of HPC originating from the biomass associated with the OP, ST and WEOM sand reactors at 25°C were 1411±33, 1343±96 and 1153±81 CFU/mL, respectively. However, the average HPC for NR water bio-acclimated sand was 866±62 CFU/mL at the same temperature, which suggests that the contribution of biodegradation to the overall DOC removal is reduced at low DOC concentration. Furthermore, ANOVA test results revealed that temperature has a significant effect on the biological activity of the sand ($p = 0.02$). It was found that the biological activity apparently increased with temperature (Table 3.4). The results indicated that the HPC values associated with the bio-acclimated sand of NR batch reactors increased by 16 and 17% at 25 and 30°C compared to the value at 20°C (723±98 CFU/mL); this illustrates that the role of biodegradation in DOC removal is increased with rising temperature. This was also confirmed with SUVA results, the SUVA values were obviously increased with temperature rise, indicating that there is preferred degradation of non-aromatic compounds. SUVA increased by 14, 16 and 21% for effluent NR waters at 20, 25 and 30°C, respectively, compared to the influent value. The same trend was observed for ST and OP waters, where SUVA increased by (34, 42 and 47%) and (12, 29 and 33%), respectively, at the same temperatures. However, WEOM water did not exhibit the same trend, its highest increase (11%) of SUVA was recorded at 20°C, and was attributed to its high humic content which is preferentially removed by adsorption (Maeng et al., 2008).

Table 3.3. Characteristics of the influent and effluent water

Influent			DOC	SUVA	NO$_3^{-1}$	NH$_3^{+1}$	PO$_4^{-3}$	pH
			mg L^{-1}	L mg^{-1} m	mg L^{-1}	mg L^{-1}	mg L^{-1}	
NR		Influent	4.05±0.69	1.82	N.D.	N.D.	N.D.	7.42
	20°C	Effluent	3.31±0.39	2.11	0.09	N.D.	N.D.	7.53
	25°C	Effluent	3.35±0.69	2.16	0.17	N.D.	N.D.	7.78
	30°C	Effluent	3.68±0.47	2.3	0.14	N.D.	N.D.	7.85
ST		Influent	10.62±0.59	1.49	0.15	2.34	1.64	7.96
	20°C	Effluent	5.72±0.34	2.27	0.64	0.25	0.71	8.11
	25°C	Effluent	5.51±0.59	2.58	0.89	N.D.	0.69	8.18
	30°C	Effluent	5.63±0.41	2.79	0.59	N.D.	0.68	8.17
OP		Influent	10.31±0.68	1.56	0.14	2.61	1.84	8.08
	20°C	Effluent	5.05±0.47	1.78	0.19	0.17	1.09	8.13
	25°C	Effluent	4.83±0.68	2.19	0.56	N.D.	1.09	8.16
	30°C	Effluent	4.65±0.39	2.34	0.56	N.D.	1.02	8.28
WEOM		Influent	9.41±0.79	2.44	N.D.	1.04	0.24	8.11
	20°C	Effluent	6.34±0.71	2.73	0.12	N.D.	0.08	8.17
	25°C	Effluent	6.75±0.21	2.29	0.36	N.D.	0.08	8.29
	30°C	Effluent	6.78±0.54	2.51	0.36	N.D.	0.07	8.24

N.D.: not detected.

Limit of quantitation (LOQ) for NO$_3^-$ (0.05mg/L), NH$_4^+$ (0.2mg/L) and PO$_4^{-3}$(0.05mg/L)

Table 3.4. HPC of batch reactors sand at different temperatures

Influent	HPC (CFU/mL)		
	20°C	25°C	30°C
NR	723.5±98	866±62	872±63
ST	1279±45	1343±96	1417±86
OP	1335±18	1411±33	1415±59
WEOM	1092±240	1153±81	1118±173
NR+HgCl₂	108±108	158±31	97±21

To analyze the organic matter characteristics of influents and effluents, samples were taken at the beginning and end of the batch experiments to measure fluorescence intensity. Three dominant peaks at definite wavelengths representing the bulk organic matter fractions (humic, fulvic and protein) were discriminated (Maeng, 2010). Apparently, fluorescence intensity of protein-like components (peak 3) decreased with increasing temperature for all organic water matrices. An exception was observed for WEOM water, which exhibited higher protein compound removal at low temperature that may be attributed to adsorption. Additionally, the results of abiotic conditions verified that the adsorption process is more influential at low temperature, even for biodegradable materials.

Table 3.5 shows that change in fluorescence intensity for each of the identified peaks is highly dependent on the organic matter composition and temperature. Humics compounds (humic and fulvic) bear a negative charge and thus tend to be removed by adsorption during the infiltration process (Abdelrady et al., 2018; Maeng, 2010). In this research, abiotic batch reactors exhibited a relatively larger decrease of humics compound fluorescence intensity at lower temperatures. Therefore, it can be concluded that humics reduction (by adsorption) exhibits an inverse relationship with temperature rise, this was also reported by Abdelrady et al. (2018). A similar trend of better attenuation at lower temperatures (20 °C) was observed for NR effluent water in which the biodegradable matter was limited. However, OP, WEOM and ST effluents showed an increase of humics fluorescence intensity. This increase displays a positive relationship with temperature rise for ST effluent and a negative relationship for OP effluents (Figure 3.1). Previous studies (Abdelrady et al., 2018; Maeng et al., 2008) suggested that bacteria present in the soil are able to form new fluorescent organic compounds associated with existing DOM during the long-term infiltration process.

Apparently, fluorescence intensity of protein-like components (peak 3) decreased with increasing temperature for all organic water matrices. An exception was observed for WEOM water, which exhibited higher protein compound removal at low temperature that may be attributed to adsorption. Additionally, the results of abiotic conditions verified that the adsorption process is more influential at low temperature, even for biodegradable materials.

The fluorescence indices were used to get insight information about the organic composition of the feed water. The results refer that ST and OP have the highest concentration of biodegradable organic matter. The fluorescence index (FI) were 1.67, 1.5, 1.36 and 1.25 for ST, OP, NR and WEOM feed waters, respectively. By contrast, WEOM feed water possessed the highest humic content, its humification index (HIX) was 0.75, while it was 0.56, 0.55 and 0.5 for ST, OP and NR, respectively.

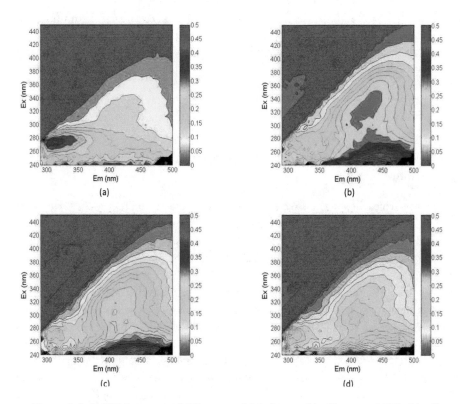

Figure 3.1. F-EEM spectra of OP water: (a) influent, (b) effluent at 20°C, (c) effluent at 25°C, (d) effluent at 30°C

Table 3.5. Fluorescence peaks intensities (RU) in oxic batch reactors operated at different temperatures using different feed water types

Feed water type	Fluorescence peaks (RU)											
	P1 (λex/λem = 245/440 nm)				P2 (λex/λem = 345-/440 nm)				P3 (λex/λem = 240/340 nm)			
	NR	ST	OP	WE OM	NR	ST	OP	WE OM	NR	ST	OP	WE OM
Inf	0.32	0.32	0.30	1.18	0.20	0.12	0.25	0.45	0.34	0.45	0.44	0.43
Eff-20 °C	0.28	0.35	0.63	1.62	0.13	0.18	0.33	0.68	0.19	0.25	0.18	0.28
Eff-25 °C	0.34	0.55	0.48	1.83	0.24	0.28	0.26	0.58	0.16	0.20	0.16	0.31
Eff-30 °C	0.37	0.61	0.35	2.18	0.24	0.34	0.2	0.54	0.12	0.18	0.16	0.34

3.3.2 Effect of feed water organic matter composition on OMPs removal

Laboratory-scale batch reactors were conducted at high temperatures (20-30°C) to assess the impact of feed water organic composition on the removal of selected OMPs. Four different water types (NR, ST, OP, and WEOM) were introduced into the batch. Biodegradation and adsorption are the main mechanisms for removal OMPs during the bank filtration processes, and they are considerably impacted by the raw water characteristics (Bertelkamp et al., 2014). In this research, a statistically significant effect of the influent organic matter composition on OMPs removal was observed at 20°C ($p = 0.02$). However, no statistically significant effect was observed at higher temperatures of 25°C ($p = 0.06$) and 30°C ($p = 0.09$), indicating that the influence of feed water organic matrix on OMP removal is lower at higher temperatures. Among the OMPs studied, removal of DDT, pyriproxyfen, pendimethalin, and picloram in NR blank reactors was 78.3±7.3%, 69.7±4.5%, 90.4±3.6%, 90.1±5.6%, and 94%, respectively. These removal efficiencies were 5 to 7% higher for the other water types tested. This implies that abiotic removal mechanisms such as volatilization and hydrolysis may play a major role in reducing the concentration of these OMPs during the batch study. Furthermore, the removal efficiency of these OMPs was higher than 99% in the batch reactors regardless of the water organic matrix, a result mainly attributed to adsorption (Figure 3.2). Similarly, β-BHC and endosulfan sulfate exhibited removal between 80 to 94% which was mainly attributed to adsorption (Figure 3.2). Thus, it can be concluded that highly hydrophobic compounds (logS >4) tend to be highly removed by (>80%) when the residence time is (≥30 days). These results are in agreement with Rodríguez-Liébana et al. (2017), who asserted that neutral hydrophobic pesticides, such as pendimethalin, are mainly removed by abiotic processes in soil irrigated with treated wastewater.

Simazine, atrazine, metolachlor and isoproturon (-2.5> logS >-4) demonstrated more persistence during the batch study (removal efficiencies varying between 16-59%) (Figure 3.2). They exhibited a relatively higher removal efficiency from ST and OP influent waters containing higher concentrations of biodegradable organic compounds. This finding implies that co-metabolism may play a role in the removal of these compounds. This is consistent with the findings of Bertelkamp et al. (2016a), who studied the effect of three feed waters with different organic fractions (hydrophilic, hydrophobic, transphilic) on the removal of OMPs (include atrazine and simazine) in three similar sand columns, and found that the biodegradability of the feed water significantly affects the removal efficiency of OMPs during the filtration process. Orlandini (1999) reported that

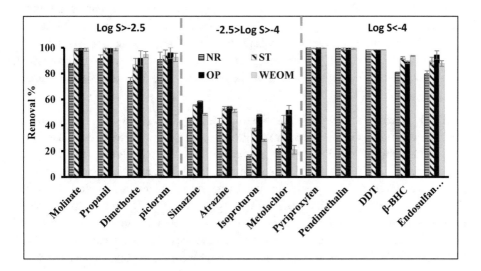

Figure 3.2. Impact of influent organic matter type on the removal of OMPs in batch reactors (20°C, oxic)

atrazine removal is increased in the presence of high concentrations of biodegradable materials which enhance the biological activity associated to the sand and enables the existing bacteria to use atrazine as a source of carbon or nitrogen. On the other hand, these moderate hydrophobic OMPs exhibited lower removal efficiencies for NR feed water that mainly ascribed to its low organic content and thereby low diversity of microorganisms in the sand. Likewise, lower removal efficiencies were observed for WEOM feed water that might attribute to its high humic content and the competition between OMPs and humic compounds for adsorption sites that should be considered during the bank filtration installing processes.

Molinate, dimethoate and propanil compounds with relatively high solubility (logS >-2.5) exhibited more biodegradable character during the batch study (Figure 3.2). According to Singh et al. (2006), highly soluble pesticides are more subject to degradation. The high biodegradability characteristics of these compounds are mainly ascribed to the presence of electron donating functional groups (amine groups) in their structures (Tran et al., 2013). Therefore, the removal efficiencies of these OMPs decreased significantly under abiotic conditions (Figure 3.3, Figure S3.1, and Figure S3.2). These results emphasize the importance of the biodegradation process in the removal of these hydrophilic compounds during the BF process. This is in agreement with other researchers (Bertelkamp et al., 2016a; Lopes et al., 2013) who discovered that adsorption of these compounds to the soil is very weak. With respect to the effect of influent organic matter, the removal efficiencies of these compounds decreased slightly under conditions of limited organic content at a temperature of 20°C (Figure 3.2). The removal of molinate, propanil and dimethoate were decreased by 12±0.65%, 9±0.52% and 13-18%, respectively, for NR during a residence time of 30 days in the batch reactors compared to their removal in water of higher influent biodegradable matter (OP, ST). Nunes et al. (2013) illustrated that co-metabolic processes mainly dominated their degradation; thus, the presence of an additional carbon source could enhance degradation.

PAH compounds are highly toxic and classified as probable human carcinogens. Blank batch reactors exhibited limited removal of these hydrophobic compounds (less than 5%) at all tested temperatures. Furthermore, no significant effect of influent organic matrix was observed on the removal of these compounds. In this research, the removal of these compounds exceeded 80% under different influent organic conditions. Lamichhane et al. (2016) demonstrated that biodegradation and hydrolysis of PAH compounds are very limited. Accordingly, it can be concluded that the high removal observed in this study is mainly attributed to sorption and prolonged residence time (30 days). It was observed that PAH compounds with more than three aromatic rings were thoroughly removed (> 99%) (Figure 3.4).

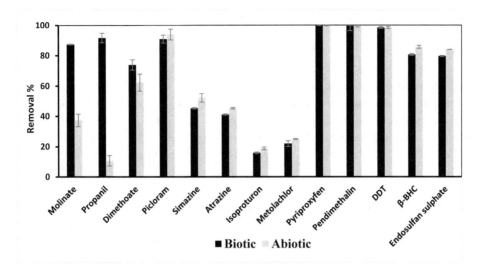

Figure 3.3. Removal of OMPs under biotic and abiotic conditions at 20°C for NR influent water

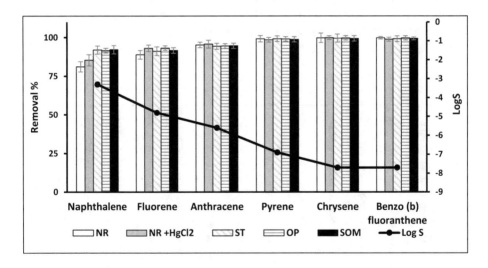

Figure 3.4. Relationship between removal of PAHs in different feed water types and LogS (at 30°C, oxic)

3.3.3 Effect of temperature on OMPs removal

To assess the effect of temperature on the removal of OMPs during the BF process, batch experiments were conducted at three different temperatures, 20, 25, and 30±2°C, using different water types in oxic conditions. Attenuation of 5 of the 13 selected pesticides investigated in this study showed temperature dependency (Figure 3.5 and Figure S3.3). The removals of pendimethalin, pyriproxyfen and DDT exceeded 95% at all three temperatures tested, although previous studies (Cheng et al., 2014; Kpagh et al., 2016; Sullivan et al., 2008) illustrated the temperature-dependent attenuation behaviour of these OMPs during the infiltration process. The high attenuation of these compounds is mainly attributed to their high adsorption efficiencies and prolonged residence times, as previously described. In the same regard, Maeng (2010) conducted laboratory-scale batch studies at 16-17°C to assess the removal of selected OMPs (e.g., pharmaceuticals) and reported as well the high removal (>80%) of the hydrophobic compounds (e.g., bezafibratem ibuprofen, naproxen, logS<-4). The same trend was observed for PAH compounds. However, naphthalene, fluorene, and anthracene exhibited lower removal values at higher temperatures (30°C), likely due to their lower adsorption efficiency at higher temperature or the desorption process which may take place during the prolonged residence time. Hiller et al. (2008) illustrated that less hydrophobic PAHs (e.g., naphthalene) demonstrate lower adsorption efficiency and higher desorption characteristics at high temperatures. Thus, post-treatment of bank filtrate may be required to remove these compounds at high temperature. Likewise, picloram exhibited a high

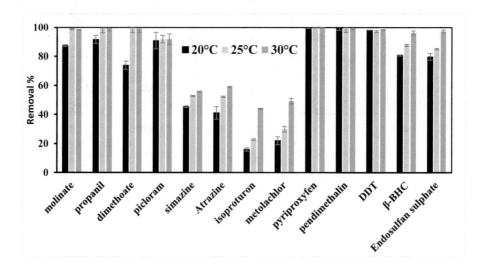

Figure 3.5. Removal of OMPs at different temperatures (20, 25, 30°C) under oxic condition for NR water

removal efficiency (> 90%) at all processed temperature, an effect that may be ascribed to abiotic process (e.g., hydrolysis and volatilization).

The temperature-dependent removal of molinate, dimethoate, and propanil was observed when the influent organic content was limited (NR), as their removal efficiencies were increased considerably at high temperature. Removal of molinate and dimethoate in abiotic batch reactors significantly increased with increase in temperature, indicating that the hydrophobicity of these compounds increased at high temperature; this was also previously reported by Rani et al. (2014). However, propanil exhibited temperature-independent removal under abiotic conditions. This implies that the biotic process plays a positive role in its removal at higher temperature.

Attenuation of atrazine, simazine, metolachlor, isoproturon, endosulfan sulfate and β-BHC were markedly increased at high temperatures for all influent waters under both biotic and abiotic conditions. The high attenuation of these hydrophobic compounds with increasing temperature is mainly attributed to their increasing adsorption capacity at higher temperature.

3.3.4 Effect of redox on removal of OMPs

The effect of three redox conditions including oxic, anoxic (NO_3^- as electron acceptor) and anaerobic (Fe^{+2}, Mn^{+2} as electron acceptors) on the removal of 19 OMPs was examined for different water characteristics at high temperature (25°C.). Effluent dissolved oxygen in the suboxic batch reactors was less than 0.2 mg/L, and the average removal of nitrate in anoxic reactors was 92±5.8%. Figure 3.6 shows the average removal of the selected pesticides at 25°C under different redox conditions.

Various studies (Bertelkamp et al., 2016b; Sullivan et al., 2008) asserted that the removal efficiencies of DDT, pyriproxyfen and, pendimethalin decreased under sub-oxic conditions. However, it was obvious in this research that the removals of these OMPs are redox-independent and exceed 99% in all batch reactors operating at 25°C. This is mainly attributed to their high adsorption efficiency at higher temperature as well as the prolonged residence time in the soil. In the same manner, PAHs also exhibited high removal efficiencies (> 99%) at the same temperature regardless of the redox conditions, as adsorption was the dominant mechanism. Apparently, if the hydrophobic compound exhibits logP > 4, it tends to be highly removed (>80%) during the filtration process (residence time: 30 days) regardless of the redox, organic composition, and temperature environment, at least for the OMPs tested during this research. Likewise, dimethoate displayed high removal efficiency (> 90%) independent of the redox condition. Inversely, Bertelkamp et al. (2016b) reported that dimethoate removal is highly reduced under sub-oxic conditions: its biodegradation rate is 72% lower under anaerobic conditions and reduced by 85% under anoxic conditions when compared to oxic conditions. This is

59

probably because the experiments were conducted at low temperature (12°C). This implies that dimethoate is preferentially removed due to an increase in microbial activity with increasing temperature regardless of the environment of redox conditions. Similarly, abiotic removal mechanisms (i.e., hydrolysis) may also have played a role in the removal efficiency at high temperature (Van Scoy et al., 2016). On the other hand, triazines (atrazine and simazine) demonstrated the same persistence under the redox conditions tested. In contrast, Gimbel et al. (2006) and Shahgholi et al. (2014) revealed that the suboxic condition resulted in higher biodegradation removal of triazines than did the oxic condition during the BF process. This contradiction can be explained in terms of the absence of organisms that are able to consume triazine compounds in the batch reactors (Bertelkamp et al., 2016b). This trend was also observed for metolachlor, which showed high persistence independent of the redox conditions. Low biodegradation of metolachlor under different redox conditions was reported in other studies (Sanyal et al., 2000). Seybold et al. (2001) studied metolachlor degradation in wetland soil and water microcosms and stated that metolachlor is degraded under anaerobic condition with a half-life of 67 days. Therefore, it can be concluded that BF does not effectively remove all of OMPs in an acceptable manner in temperate regions and that post-treatment may be required for drinking water purposes.

Molinate and propanil exhibited a high dependency on the redox conditions, since their biodegradation was highly increased under oxic conditions compared to anoxic and anaerobic conditions. The removal of these pesticides under different redox conditions during BF has not yet been studied. However, these results are consistent with Lopes et

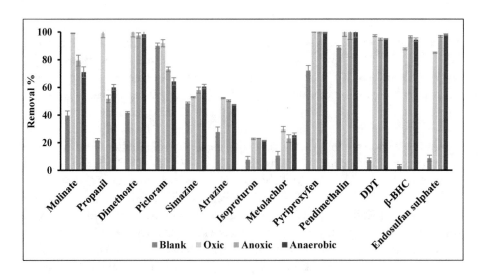

Figure 3.6. Removal of OMPs under different redox conditions at 25°C for NR influent

al. (2013) and Nunes et al. (2013) who tracked the environmental fate of these hydrophilic compounds in wet soils. Therefore, it can be concluded that the compounds require longer residence time to be removed during the suboxic infiltration process. However, some molinate transformation products such as molinate sulfoxide are also potential toxicants and have higher solubility compared to the original compound; therefore, they have high potential to be leached during bank filtration, which requires further investigation.

Endosulfan sulfate and β-BHC exhibited relatively higher removal under sub-oxic conditions. The removal efficiencies for endosulfan sulfate and β-BHC were increased by 8.7 and 11.8% for the anoxic condition and 6.7 and 13% under anaerobic conditions, respectively. Baczynski et al. (2010) illustrated that organochlorine pesticides are degraded effectively under anaerobic conditions and that this process is temperature-dependent, increasing slightly at higher temperature.

3.4 CONCLUSIONS

Batch studies were conducted to assess the removal efficiency of some micro-organics pollutants during the BF process (travel time =30 days) under different environmental conditions. Based on the results of this research, the following conclusions can be drawn:

- Labile compounds (i.e., protein-like) are the most amenable DOM fraction to be removed during BF at all the tested temperatures.

- Atrazine, simazine, isoproturon, and metolachlor (-2.5> logS >-4) were the most persistent compounds and were mainly removed by adsorption process (<45%, residence time =30 days). The removal of these compounds was enhanced at higher temperatures and in the presence of organic matter. This finding indicates that co-adsorption, as well as co-metabolism, may play significant roles in the removal of these compounds.

- Poorly soluble OMPs with logs<-4 (i.e., DDT, pyriproxyfen, pendimethalin, β-BHC, and endosulfan sulfate) were highly removed (>80%) by adsorption (residence time =30 days) regardless of the environmental conditions (i.e., temperature, redox and influent organic characteristics). However, these compounds may degrade in the soil and produce more toxic and more soluble compounds which may leach into the bank filtrate. Further investigation is required in this phenomenon.

- The removal efficiencies of soluble OMPs with logS>-2.5 (molinate, propanil, and dimethoate) were below 40% under abiotic conditions. Removal efficiencies of these hydrophilic compounds increased to >70% under biotic conditions. This implies that biodegradation has a key role in the attenuation of these compounds during BF. Thus, influent temperature and biodegradable organic matter concentration may affect the removal of these OMPs due to their influence on microbial activity and the co-metabolism processes taking place in the filtration area. The removal of these compounds was found to be higher than 95% at 25°C.

- DDT, pyriproxyfen, and pendimethalin exhibited redox-independent behaviour at high temperatures (25°C), as their removal efficiencies exceeded 95%.

- Molinate and propanil showed redox-dependent removal behaviour with high attenuation under oxic conditions (>87%). In contrast, β-BHC, and endosulfan sulfate showed slightly higher biodegradation under sub-oxic conditions (>94%).

- Atrazine, simazine, isoproturon, and metolachlor, characterized by persistence properties under oxic conditions, were also not removed under sub-oxic conditions. Thus, post-treatment may be required to remove these OMPs from bank filtrate.

- Polyaromatic hydrocarbons (PAHs) compounds exhibited high removal (>80%) regardless of the organic composition of the feed water and the redox environment. However, a significant effect of temperature was observed for lower hydrophobic PAHs (e.g., naphthalene, fluorene, and anthracene) at high temperature that may attribute to lower adsorption efficiency or desorption process, which needs to be investigated

In summary, the results indicated that higher temperature enhanced the removal of OMPs during the BF, although it decreased the attenuation of bulk organic matter.

3.5 SUPPLEMENTARY DOCUMENTS

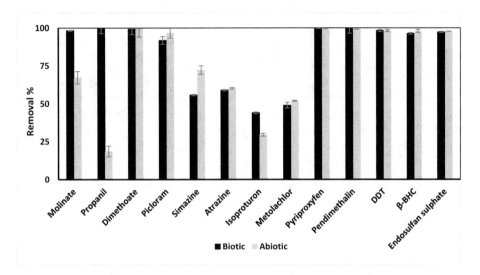

Figure. S3.1 Effect of biotic/abiotic conditions on the removal of OMPs at 30 °C

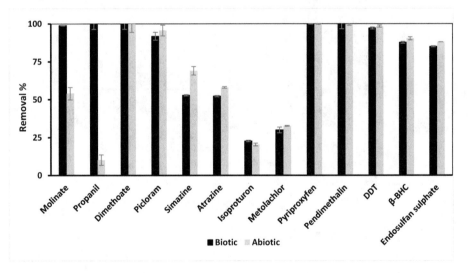

Figure S3.2 Effect of biotic/abiotic conditions on the removal of OMPs at 25 °C

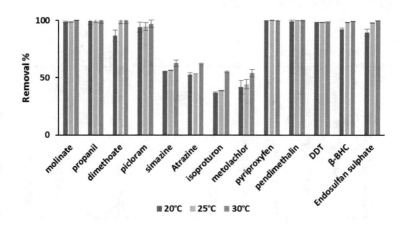

(a)

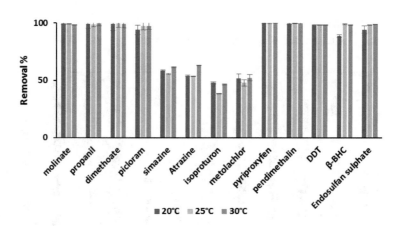

(b)

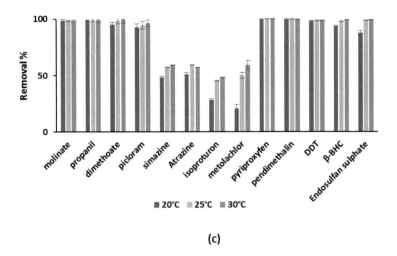

(c)

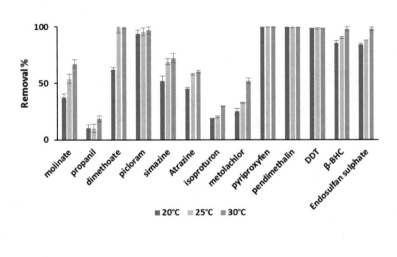

(d)

Figure S3.3. Effect of temperature on OMPs removal using different water types: (a) ST, (b) OP, (c) WEOM, (d) NR+HgCl₂

4

REMOVAL OF METALS DURING BANK FILTRATION

ABSTRACT

The effectiveness of bank filtration (BF) is highly dependent on the source water quality (e.g., the organic matter composition, pH, temperature, and concentration of heavy metals (HMs)) and hydrogeological conditions at the subsurface. In this study, the impact of dissolved organic matter (DOM) on the removal of selected metals (Cu, Zn, Pb, Se, and Ni) during BF was investigated. Laboratory-scale column studies were conducted at 30 °C with feed water sources of different organic matter composition. The selected HMs were spiked at concentrations similar to their levels in surface waters. Fluorescence excitation–emission matrix coupled with parallel factor analysis (PARAFAC–EEM) was used to characterise the organic composition of the different feed waters. Moreover, another series of column studies was conducted to assess the impact of DOM type (humic, protein) and concentration on the removal of HMs. The experimental results revealed a high Pb(II) removal efficiency (> 90%) during filtration, which depends only slightly on the organic matter content of the feed water. In contrast, Cu(II), Zn(II), and Ni(II) exhibited relatively lower removal efficiencies (65%–95%) for the columns fed with different water sources. These removal efficiencies decreased considerably in the presence of humic compounds, regardless of their origin (terrestrial or microbial). On the other hand, the removal efficiency of Se(IV) increased as the concentration of biodegradable organic matter in the feed water increased, and a positive correlation was found between the Se(IV) removal efficiency and the protein-like fluorescence intensity ($\rho = 0.72$, $p < 0.01$). In general, it can be concluded that the organic composition of the source water affects profoundly the removal and sorption of HMs during the BF, and should be considered in the design of bank filtration systems.

4.1 INTRODUCTION

Heavy metals (HMs) are widely distributed in the earth's crust and are naturally non-biodegradable. A small number of HMs (in trace amounts) have an essential role in the metabolism of humans and animals. However, higher concentrations may be toxic. Metals such as mercury (Hg), cadmium (Cd), arsenic (As), chromium (Cr), thallium (Tl), zinc (Zn), nickel (Ni), copper (Cu), selenium (Se), and lead (Pb) are commonly found in surface waters. These metals constitute a significant public health concern because of their bio-accumulative nature and must therefore be removed from drinking water (Martin et al., 2009). Excess exposure to Pb causes irreversible brain damage and encephalopathy symptoms. Drinking water with high levels of Cu may cause nausea, vomiting, stomach cramps, and sometimes diarrhoea (Duruibe et al., 2007). In addition, high Se concentration can cause numbness in the extremities, circulation problems, neurological impairment, and fingernail or hair loss. High levels of Zn in drinking water may cause stomach cramps, vomiting, and sometimes nausea (Martin et al., 2009). Moreover, a high concentration of Ni can damage the DNA of the hands (Sharma, 2014).

Removal of HMs during BF is highly dependent on the site conditions, source water quality matrix, and the type of the HM e. Various studies have been conducted to investigate the biogeochemical processes taking place during the infiltration of river water into alluvial aquifers and its impacts on the removal of HMs (Bourg et al., 1993). These studies highlighted that the adsorption of HMs during the filtration process is considerably impacted by the water quality matrix of the raw water, more specifically, the pH, dissolved oxygen, and dissolved organic matter (DOM). DOM contains multiple functional groups that have an affinity for HMs and hence may affect the mobility of HMs in aquatic and soil systems during the BF process (Weng et al., 2002).

The interaction of DOM and HMs in surface water occurs in different ways (Weng et al., 2002). This affects the chemical state and availability of HMs in natural water systems. DOM in surface water forms complexes with HMs, which reduces bioavailability and toxicity to aquatic organisms (Sharma, 2014). The type of DOM such as humic substances (fulvic acid and humic acid) influences the geochemical mobility of metal ions during the infiltration (Weng et al., 2002). According to Reuter et al. (1977), metal–humic complexes in natural water systems have a higher stability than inorganic–metal compounds. Hence, the presence of DOM affects the variation in the behaviour of metal pollutants in natural water systems.

The risk of HM contamination of drinking water is increasing, in particular in developing countries where industrial and human activities result in the discharge of waste into water bodies (Sharma, 2014). Consequently, this study focuses on the influence of DOM composition of the feed water on the removal of HMs during the BF process, particularly in regions with hot climates. Therefore, laboratory-scale column studies were conducted

69

at a controlled room temperature (30 °C) to investigate the effect of the feed water source and natural organic matter (NOM) type (humic-like, protein-like tyrosine, the combination of protein-like and humic acid) on the removal of HMs (Cu, Zn, Ni, Pb, and Se) during the infiltration process.

4.2 MATERIALS AND METHODS

4.2.1 Filter media characteristics

The silica sand (grain size 0.8-1.25 mm) used in this study was bought from Wildkamp company (Netherlands). The characteristics of the sand is presented in Table 2.1. The media was washed with non-chlorinated tap water to discard the debris before introducing into the columns. The average bulk density and porosity of sand were 1500 kg/m³ and 0.42, respectively. The organic content and the HMs concentrations of the sand were determined using the acid-extraction method described in (Sastre et al., 2002). The biological activity associated with the sand was estimated in terms of Adenosine triphosphate (ATP) following the technique described in previous studies (Abdelrady et al., 2018; Abushaban et al., 2017).

4.2.2 Column experiment

Laboratory-scale soil column setup was used to assess the impact of organic matter on the removal of HMs during the filtration process. The PVC pipe columns with an internal diameter of 4.2 cm and a total height of 50 cm were used for all experiments. Silica sand was filled up on top of the gravel for about 40 cm in all columns. Then the columns were fitted with connectors from both ends to fix the influent and effluent tubes. The influent tube was connected to the influent tank (10 litres) by using a polyethylene tube.

The influents and effluents tubes were replaced after every five days and also covered with black materials and foils to reduce biofilm formation during the experimental process. The influent tank was cleaned with distilled water and if necessary with HCl before introducing the feedwater into it. Aeration procedure in the feed water tank was implemented to ensure the experiment was performed under aerobic conditions. The dissolved oxygen (DO) was monitored continuously to ensure the oxygen supplied is enough to support the experimental condition (DO>7 mg/L). The experiments were conducted at a temperature-controlled room (30 °C) to analyse the possibility of removing HMs during BF in hot climate areas. Different parameters were monitored periodically to ensure that the experiments were conducted under the desired conditions. Such parameters included the dissolved oxygen, pH of the influent, infiltration rate and the concentration of HMs in the feed water.

Effect of DOM composition of raw-water on HM removal during BF

Four columns were fed with different types of source waters to investigate the effect of the feed-water organic matter composition on the removal of HMs. Therefore, four types of feed water were prepared from different sources including (i) non-chlorinated tap water (NCTW) with a low organic content, (ii) Delft canal (DC) water collected from a canal, (iii) secondary treated wastewater obtained from the Harnaschpolder WWTP and mixed with Delft canal water to simulate contaminated surface water systems (DCWW), and (iv) water containing extractable organic matter (WEOM) which contains higher concentrations of humic compounds, which was prepared following the method proposed by Guigue et al. (2014). The physical–chemical water quality parameters of all feed waters were analysed before the start of the experiment (Table 4.1). The pH of the feed waters was adjusted to 7.8±1. The experiment was conducted at different hydraulic rates (0.3, 0.6 and 1.0 m/day).

Table 4.1. Characteristics of the feed waters used for the laboratory- columns experiment

Parameters	DC	DCWW	WEOM	NCTW
pH	8.3 ± 0.20	8.0 ± 0.10	8.5±0.10	7.8 ± 0.20
DO (mg/L)	8.4 ± 0.20	7.9 ± 0.20	7.9±0.10	8.5 ± 0.30
EC (µS/cm)	881 ± 0.58	838 ± 0.80	700±0.50	551 ± 0.58
Total Nitrogen (mg/L)	1.1 ± 0.06	6.7 ± 0.14	1.8±0.01	1.1 ± 0.06
NH_4^- -N (mg/L)	0.03 ± 0.00	1.2 ± 0.03	0.7±0.02	0.03
PO_4-P	< 0.3	1.6 ± 0.00	< 0.3	< 0.3
Se(IV) (µg/L)	< 2	< 2	< 2	< 2
Cu(II) (µg/L)	16.7 ± 0.28	8.2 ± 2.30	138.5±2	5.2 ± 2.30
Pb(II) (µg/L)	< 5	< 5	< 5	< 5
Ni(II) (µg/L)	< 5	< 5	< 5	< 5
Zn(II) (µg/L)	29.5 ± 2.10	27.5 ± 0.71	47.5±0.5	7.1 ± 1.30

All values were presented as Mean value ± standard deviation, n=7

Effect of NOM type on HM removal during BF

This experiment was conducted to investigate the impact of the NOM type and concentration on the HM removal efficiency during a BF. Therefore, four columns were developed and ripened at 30 °C, and hydraulic rate of 0.3 m/day. The first column was fed with NCTW and used as a control sample. The other columns were fed with mixtures of NCTW and different NOM types: (i) humic (HA), (ii) TY, and (iii) a mixture of 50% HA and 50% TY (HA:TY). Each type of NOM was injected at four different concentrations (5, 10, 15, and 20 mg-C/L). The experiments were conducted for each concentration and for the residence time of seven days, during which influent and effluent samples were taken daily.

Effect of NOM type on specific HM removal efficiency

The main target of this experiment was to analyse the impact of NOM type on the removal efficiency of specific HM's during infiltration and to assess whether the presence of multiple metals in the feed water can affect the adsorption characteristics of other HM's. Therefore, eight columns were prepared and ripened for two months. Subsequently, the columns were fed with NCTW containing different NOM types. Cu(II) and Se(IV) were spiked individually into the feed water of the four columns, and the experiment took seven days.

The Thomas model was used to estimate the breakthrough of the metals and impact of the NOM type on the adsorption parameters (adsorption capacity and rate) of the tested metals during the filtration process. This kinetic model assumes that adsorption is governed by the mass transfer and chemical reaction processes, and this reaction is a reversible second-order reaction, which obeys the Langmuir adsorption kinetics. Thomas model in the linear form is described as follows (equation 1) (Chu, 2010):

$$\ln\left(\frac{C_0}{C_t} - 1\right) = \frac{K_{TH}\, q_0\, X}{Q} - K_{TH} C_0\, t \qquad (1)$$

K_{TH} is Thomas adsorption rate constant (L/hr.µg), q_0 is the adsorption capacity (µg/g), X is the mass of adsorbent (g), Q is the flow rate of the feed water (L/hr), C_0 and C_t are the initial and breakthrough concentrations (µg/L).

4.2.3 Analytical methods

Both influent and effluent samples were collected from the columns and filtered using 0.45 µm filtration (Whatman, Dassel, Germany). The samples were acidified with 0.5 ml of concentrated HCl and stored at a controlled temperature room (4 °C). The metal (Cu, Zn, Ni, and Pb) concentrations were determined by inductively coupled plasma–mass spectrometry (Xseries II Thermo Scientific). The limit of detection (LOD) of metals was 10 µg/L. The Se concentration was quantified with a graphite furnace atomic absorption spectrophotometer (Solaar MQZe GF95, Thermo Electron Co.), where the LOD was 5 µg/L.

The combustion technique was employed to determine the organic content (DOC in mg-C/L) of the influent and effluent samples using a total organic carbon analyser (TOC-VCPN (TN), Shimadzu, Japan) (LOD = 20 mg/L). Specific ultraviolet absorbance (SUVA$_{254}$) [L mg^{-1} m^{-1}] was estimated to define the aromaticity of the influent and effluent. SUVA$_{254}$ was calculated as a ratio between the DOC of the water sample and its ultraviolet absorbance (UV) at 254 nm [m^{-1}]. The UV absorbance was determined using a UV/Vis spectrophotometer (UV-2501PC Shimadzu).

The organic constituents of the samples were determined with the fluorescence excitation–emission matrix (EEM) technique (Fluoromax-3 spectrofluorometer, HORIBA Jobin Yvon, Edison). The fluorescence intensity of the samples was determined at the excitation wavelengths (λex) 240–452 nm with 4 nm intervals and at the emission wavelengths (λem) 290–500 nm with 2 nm intervals.

The fluorescence indices, including the humification index (HIX), fluorescence index (FIX), and biological index (BIX), were used to characterise the organic fluorescence characteristics of the feed water. These indices were described in details by Gabor et al. (2014). HIX is an indicator of the humification degree of the feed water, and estimated as follows (Ohno, 2002):

$$HIX = \frac{\sum FI_{\lambda em\ (434-480\ nm)}}{\sum FI_{\lambda em\ (434-480\ nm)} + \sum FI_{\lambda em\ (300-344\ nm)}} \quad at\ \lambda ex\ (254\ nm) \tag{2}$$

FIX is used to detect the source of DOM in water (terrestrial or microbial). The value of FIX for each sample was estimated following Equation 3 (Gabor et al., 2014), and it varies between 1.2-1.8 with lower values refer to DOM of microbial origin.

$$FIX = \frac{FI_{\lambda em\ (450)}}{FI_{\lambda em\ (500)}} \quad at\ \lambda ex\ (370\ nm) \tag{3}$$

BIX was used to estimate the contribution of microbial DOM to the total organic content of the feed water. This parameter was determined using Equation 4 (Huguet et al., 2009).

$$BIX = \frac{FI_{\lambda em\ (380)}}{FI_{\lambda em\ (430)}} \quad at\ \lambda ex\ (310\ nm) \tag{4}$$

4.2.4 Fluorescence modelling

The fluorescence EEM spectroscopy was combined with the parallel factor analysis (PARAFAC) algorithm to discretise the fluorescence dataset into independent fluorescence components. This technique has been widely used to identify organic matter characteristics of surface and subsurface water systems (Abdelrady et al., 2018; Osburn et al., 2016a).

73

A fluorescence dataset of 80 samples collected from the feed water was used to develop and validate the PARAFAC models following the steps proposed by Murphy et al. (2013).

4.2.5 Data analysis

The two-way analysis of variance (ANOVA) and post-hoc Tukey test were used to assess the significance of the impact of organic composition of the feed water and NOM type on the removal of HMs during the column filtration process; the impact was considered significant if the level of significance (p) was below or equal to 0.05.

In addition, a Spearman rank correlation (ρ) analysis was conducted to determine the relationship between the fluorescence characteristics of the feed water and the HM removal during the filtration process. The strength of the relationship was considered high if the correlation value $\rho > 0.7$, moderate if $0.4 < \rho > 0.7$, and low if $\rho < 0.4$ (Hinkle et al., 1988).

4.3 RESULTS

4.3.1 Impact of feed water source on HMs removal

The effect of the organic composition of the feed water on the removal of heavy was studied in laboratory-scale columns using different water sources spiked with varying concentrations of HMs. The experiment was conducted at different hydraulic rates (0.3, 0.6 and 1.0 m/day).

Organic characteristics of the feed water

Four different feed water with different organic composition was applied to the laboratory-scale columns. The organic characteristics of the feed water were determined and summarized in Table 4.2. NCTW had the lowest concentration of organic matter; its value ranged between 3.47 and 3.67 mg-C/L. This feed water has low SUVA (1.76±0.14 L/ mg-g) and HIX (0.49) values, which infers to its aliphatic organic composition. The DOC of DC water (11.25±0.81 mg/l) and DCWW (11.41±0.59 mg/l) were significantly higher than that of WEOM feed water (9.7±0.24 mg/l). However, the WEOM contained a higher concentration of aromatic compounds than DC and DCWW; the average SUVA values for DC, DCWW and WEOM were 2.99±0.11, 3.72 ± 0.39 and 3.82 ± 0.27 L/mg-g, respectively.

These results were confirmed with the fluorescence indices results which revealed that WEOM had a higher humic concentration (HIX=0.84) and lower microbial-derived compounds (BIX=0.54) than DC (HIX=0.8, BIX=0.72) and DCWW (HIX=0.74,

BIX=0.79). Moreover, the FIX of WEOM was obviously more moderate than that of other feedwaters, which implies that its humic content is mainly originated from terrestrial sources. The PARAFAC model illustrated that the terrestrial humic compounds are the main contributors to the organic fluorescence spectrum of WEOM feed water. Whereas, DC and DCWW feedwaters contain a higher content of microbial humic and protein-like than WEOM. In contrast, NCTW retained with low content of the three PARAFAC components.

Table 4.2. Organic characteristics of the feed waters used for the laboratory- columns experiments

Parameters	Units	DC	DCWW	WEOM	NCTW
UVA$_{254}$	(cm^{-1})	0.34±0.04	0.45±0.04	0.37±0.02	0.06±0.01
SUVA	(L/mg-g)	2.99±0.11	3.72±0.39	3.82±0.27	1.76±0.14
HIX	----	0.8±0.08	0.74±0.05	0.84±0.07	0.49±0.03
FIX	----	1.23±0.27	1.37±0.19	1.06±0.23	1.04±0.33
BIX	----	0.72±0.11	0.79±0.21	0.54±0.06	0.39±0.05
PC1	RU	0.91±0.15	1.02±0.28	1.6±0.21	0.09±0.02
PC2	RU	1.06±0.33	1.2±0.25	1.05±0.28	0.14±0.08
PC3	RU	0.76±0.12	0.85±0.19	0.54±0.3	0.15±0.05

Effect of the organic composition of the feed water on HMs removal

The results showed that Pb(II) had the highest removal during the filtration process; the effluent concentration of Pb(II) for all the columns was less than the limit of detection (5 µg/L) of the instrument under all the experimental conditions. However, Cu(II), Zn(II) and Ni(II) exhibited relatively lower removals ranging between 65 to 95%. These removals were significantly dependent ($p < 0.05$) on the organic concentration and composition of the feed water, with a preferential removal at a low organic concentration (Figure 4.1, S4.1 and S4.2). The removals of Cu(II), Ni(II) and Zn(II) for DC feed water were 87±1, 90±4, 91±1%, for DCWW 61±4, 77±6, 71±2 %, for WEOM 76±2, 85±3, 87±3% and for NCTW 99±0.2, 93±2, 97±0.3% respectively at hydraulic loading rate of 0.3 m/day. These removals decreased by 1-2% at higher hydraulic rates (0.6, 1.0 m/day). Statistical analysis revealed that there is no significant ($p = 0.2\text{-}0.47$) effect of hydraulic

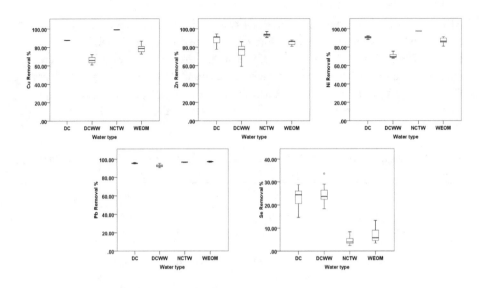

Figure 4.1. Removal of Cu, Zn, Ni, Pb (initial concentration =150 µg/L) and Se (10 µg/L) during aerobic column infiltration at HLR = 1.0 m/day (temperature=30 °C).

loading rate on the removals of these metals (Cu(II), Zn(II), Ni(II)) during the filtration process. Nevertheless, Se(IV) was the most persistence metal to be removed during the filtration process. The removal of Se(IV) was highly dependent on the feed water organic composition ($p<0.001$). The average removals of Se(IV) for the columns fed with DC, DCWW, WEOM and NCTW were 28±9, 34±8, 7±3 and 4±1% respectively at hydraulic loading rate of 0.3 m/day. This removal decreased by 2-8% when the hydraulic rate increased to 1 m/day. However, there was no significant effect ($p =0.2$) of infiltration rate on the removal of Se(IV). The removal of Se(IV) was found to be in a positive relationship with the biological activity (ATP concentration) associated to the sand. The ATP concentrations of media for the column fed with DCWW, DC, WEOM, and NCTW were 9.3±1.28, 9.2±1.04, 5.1±0.33, 3.3±0.47 µg/g, respectively.

4.3.2 Impact of fluorescence organic compounds on HMs removal

Fluorescence modelling

The PARAFAC-EEM model technique successfully decomposed the fluorescence data collected from the influents of the columns into three main components representing different organic composition groups. The model was split-half validated and explained 99.7% of the dataset variability. The contour plots and loadings of the identified fluorescence components are presented in Figure 4.2. Openfluor database, which uses

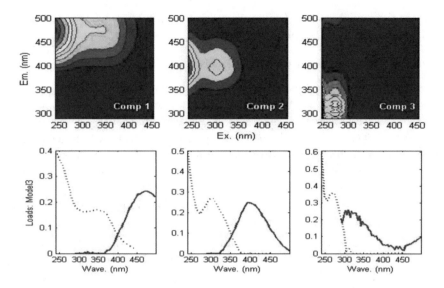

Figure 4.2. Contour plots of the three PARAFAC components separated from the complete measured F-EEMs dataset for the influent and effluent of the columns experiment

tucker's congruence coefficient (TCC) to determine the similarity degree between the components, was used to identify the PARFAC components separated and matched them to those recognised at the same excitation and emission wavelengths in previous studies (Table S4.1). The three PARAFAC components showed multiple excitations. According to Lu et al. (2009), the fluorescence component, which contains several maximum excitation wavelengths and has the same emission characteristics, are derived from the same organic fluorophores group. In this research, the first component (PC1) showed maximum excitation wavelengths (λex) at 240 and 344 nm and maximum emission (λem) at 474 nm. This component is best corresponding with humic compounds derived from terrestrial sources (Kulkarni et al., 2018; Shutova et al., 2014). These compounds are prevalent in surface water systems and characterized by condensed structure (Abdelrady et al., 2018). Component 2 (PC2) exhibited two maximum λex at 240 and 300 nm, and a single maximum λem peak at 396 nm. This component is well-matched with humic compounds originated from microbial sources and characterizes with moderate molecular weight (ranges between 650 and 1000 Da) (Kothawala et al., 2014; Osburn et al., 2016a). Component 3 (PC3) displayed two maximum λex peaks at $\leq$240 and 272 nm and maximum λem peak at 316 nm. This component closely resembles protein-like fluorophore, which mainly encompasses tyrosine and tryptophan compounds originated from microbial sources (Gonçalves-Araujo et al., 2016; Yang et al., 2019). These compounds have high biodegradability characters, and therefore, they are highly removed during BF process (Abdelrady et al., 2019).

Relationship between DOM fluorescence composition and HMs removal

The correlations between HMs removals and organic fluorescence characteristics of the feed water are presented in Figure 4.3. The results demonstrated that the organic matter concentration and fluorescence composition of the feed water have a significant impact on HMs behaviour during the filtration process. Pb(II) was the least impacted metal by the variations of the organic concentration of the feed water; the correlation between Pb(II) removal and organic matter concentration was (ρ =-0.21). Conversely, DOM was found to suppress the adsorption efficiency of Cu(II), Ni(II) and Zn(II) into the sand surface, negative relationships were observed between the removals of these metals and organic concentration of the feed water. The fluorescence indices results revealed that the humic content of the feed water affects the adsorption of Cu(II), Zn(II) and Ni(II) negatively, negative correlations were found between the removal of the mentioned metals and the fluorescence characteristics (HIX, BIX and FIX) of the feed water. Cu(II) exhibited relatively higher adsorption potential which was affected by the change in the humic content of the feed water than Ni(II) and Zn(II), the correlations between Cu(II), Ni(II) and Zn(II) removals with HIX of the feed water were -0.63, -0.53 and -0.36, respectively. The PARAFAC data revealed that terrestrial humic and microbial humic both are highly correlated with the removal of Cu(II), Ni(II) and Zn(II), which implies that humic

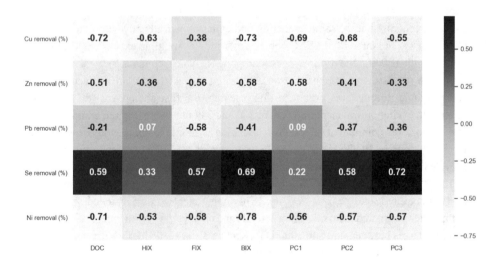

Figure 4.3. Correlations between the metals removals (%) and the fluorescence characteristics of the feed water

compounds, regardless of their structures, can react with the three metals and reduce their adsorption efficiency.

The organic matter, in contrast, was observed to enhance the removal of Se(IV) during filtration, a strong positive correlation was detected between Se(IV) removal and organic concentration of the feed water. The fluorescence data revealed that Se(IV) removal weakly relies on the level of humic compounds in the feed water, a weak correlation (ρ =0.33) was detected between Se(IV) removal and HIX index of the feed water. However, a high positive relationship was detected with BIX (ρ =0.69) and FIX (ρ =0.57) indices. Furthermore, it was noted that Se(IV) removal is highly correlated with the fluorescence intensity of the protein-like (ρ =0.72) and microbial humic compounds (ρ =0.58). Therefore, it can be concluded that Se(IV) removal during the filtration process is more linked to biodegradable matter concentration of the feed water.

4.3.3 Impact of NOM on HMs removal

Laboratory-scale column experiments were conducted to have a better understanding of the effect of organic matter on the removal of HMs during filtration using different NOM (HA, TY and a mixture of HA: TY) at different concentrations (5, 10, 15, 20 mg/L) in feed water.

Effect of NOM type and concentration on HMs removal

Three peaks were identified in the influents and effluents of the column based on their maximum fluorescence intensity at discrete excitation and emission wavelengths (Baghoth et al., 2011; Coble et al., 2014). The spectrum fluorescence matrix of the NCTW spiked with humic displayed two peaks; the first peak (P1, primary humic) was identified at maximum excitation wavelengths between 240-250 nm and maximum emission wavelengths between 380-460 nm. The second humic peak (P2, secondary humic) exhibited maximum fluorescence intensity at excitation and emission wavelengths between 280-320 nm and 400-450 nm, respectively (Coble, 1996; Leenheer et al., 2003), which implies that the spiked humic material was composed of two fluorescent humic compounds. In case of NCTW spiked with tyrosine solution, the fluorescence matrix showed one peak (P3) at excitation wavelengths of 250-280 nm and emission wavelengths of 280-350 nm. The NCTW spiked with the mixture (humic and tyrosine) showed three peaks (P1, P2, and P3) at the same wavelengths mentioned above (Figure S4.1). The organic fluorescence characteristics of each sample is presented in Table 4.3.

Table 4.3. Peaks intensity of the representative influent samples of NOM types from F-EEM contour maps

	Maximum peak intensity (RU)		
	Primary humic (P1)	**Secondary humic (P2)**	**Protein- like (P3)**
NCTW + HA	1.07	1.2	0.19
NCTW + TY	0.08	0.05	1.07
NCTW + (HA: TY)	1.13	0.6	0.75

The type and concentration of NOM have a significant influence on the removal of metals during the filtration process (Figure 4.4). The removal of Pb(II) exceeded 90% in the presence of tyrosine. The Post Hoc Tests (Turkey method) showed that there are no statistically significant differences in the removal of Pb(II) from feed water containing different concentrations of TY or a mixture of HA and TY. This points to the remarkable role of labile organic compounds (TY) in enhancing the removal of Pb(II) during the filtration process. However, for the column fed with HA, the Pb(II) removal exhibited a gradually decreasing trend with increasing HA concentration in the feed water. Pb(II) removal was 97 ± 0.12, 93 ± 1.5, 86 ± 2.5 and $73\pm3\%$ when the feed water humic concentration was 5, 10, 15, and 20 mg-C/L, respectively. This implies that HA compounds at high concentration (> 5 mg-C/L) have enough capacity to suppress the adsorption of Pb(II) onto the sand surface.

The type of NOM significantly affects the removal of Cu(II), Ni(II) and Zn(II) during the filtration process, with a preferential removal in the presence of TY. However, no significant difference in the removal of these metals was observed for the column fed with different concentrations of TY as their removal efficiencies exceeding 90%. HA compounds, in contrast, were found to decrease the adsorption efficiencies of these metals during the filtration process. The removal efficiencies of Cu(II), Zn(II) and Ni(II) were 85 ± 5, 91 ± 2 and $96\pm0.6\%$ for the column fed with a 5 mg-C/L of HA, respectively. These removals reduced to 19 ± 8, 66 ± 5 and 68 ± 5 when the concentration of HA of the feed water was 20 mg-C/L, respectively. This indicates that Cu(II) is the most influenced by the increasing of HA concentration in the feed water. The same trend was noted for the column fed with NCTW spiked with HA and TY mixture, where the removal of Cu(II), Ni(II) and Zn(II) were reduced by 45, 18 and 12% when the feed (HA:TY) mixture concentration increased by 15 mg/L. Thus, it can be concluded that humic compounds

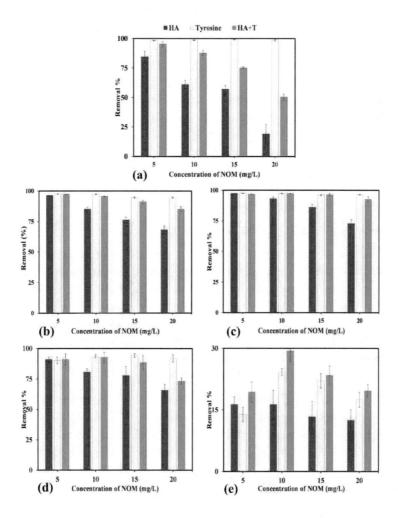

Figure 4.4. Removal of (a) Copper, (b) Zinc, (c) Lead, (d) Nickel and (e) Selenium during the column infiltration using NCTW spiked with different NOM concentrations (5, 10, 15, 20 mg-C/L) at HLR = 0.3 m/day

have a higher capacity to suppress the adsorption efficiencies of Cu(II), Ni(II) and Zn(II) onto the sand surface than TY compounds.

The removal of Se(IV) during filtration was highly dependent on the type and concentration of NOM in the feed water. The column fed with a mixture of HA and TY exhibited the highest capacity to remove Se(IV) during filtration. The removal of Se(IV) was ranged between 27-33% for the column fed with NCTW spiked with 10 mg-C/L of

81

HA:TY mixture. These removals decreased by 7-10% for the column fed with 20 mg-C/l of the mixture. Columns fed with TY exhibited a relatively lower capacity to remove Se(IV). Se (IV) removals were 14±2, 25±0.8, 22±2 and 18±2% for the columns fed with 5, 10, 15 and 20 mg-C/L of TY respectively. The results of the post hoc Tukey's test revealed that there no significant difference in the removal of Se(IV) between the columns fed with TY and HA:TY mixture. However, Se(IV) exhibited higher persistence to be removed in the absence of TY, the removal of Se(IV) was 16±3% for the column fed with 10 mg-C/L of HA. This removal was reduced to 13±2 and 12±1% when the feed water contains 15 and 20 mg-C/L of HA, respectively. Therefore, it can be concluded that the biodegradable and low molecular weight compounds are the motivated organic compounds to remove Se(IV) during the filtration process.

Kinetic study of Cu(II) and Se(IV) during the infiltration process

Among the tested metals, Cu(II) and Se(IV) were the most impacted ones by the variations of NOM type and concentration during the filtration process (as observed in the previous section). Therefore, laboratory-scale column studies were conducted to gain insight into the kinetics of adsorption of these two metals under different experimental conditions and to determine the impact of the presence of other metals. This set of experiments were conducted under the same experimental condition as earlier; however, a fixed concentration of each metal was injected individually into the NCTW feed water spiked with different NOM types (HA, TY, HA:TY).

Thomas model was used to describe the breakthrough curves and estimate the adsorption parameters (adsorption capacity and adsorption rate) of Cu(II) and Se(IV) during the filtration process (However, this effect was negligible for the columns fed with NCTW and TY.

The results showed that the experimental data fit the model very well (high R^2 and low X^2) (Figure S4.2). A significant difference in the adsorption behaviour of Cu(II) for the column fed with different NOM composition was observed (However, this effect was negligible for the columns fed with NCTW and TY.

Tyrosine was found to have a positive effect on the adsorption characteristics of Cu(II) during the filtration process. The breakthrough time of Cu(II) was longer for the column fed with NCTW and TY than the control column fed with only NCTW. Based on the Thomas model, the adsorption capacity of Cu(II) increased from 2.01 to 2.89 µg/g when 10 mg-C/L of TY was added to the feed water (Table 4.4). In contrast, HA compounds exhibited a negative effect on the adsorption capacity of Cu(II). Its value decreased to 0.53 and 1.01 µg/g when HA and HS:TY mixture were added to the NCTW feed water, respectively. Furthermore, the kinetic results revealed that the presence of other metals in the raw water reduces the effect of humic on Cu adsorption and increase the adsorption capacity of sand, the adsorption capacity were 2.83 and 1.9 µg/g for the columns fed with

HA and HA:TY mixture, respectively, in the presence of the other metals. However, this effect was negligible for the columns fed with NCTW and TY.

For Se(IV) (Table 4.5), the adsorption capacity of sand was lower for the column fed with HA (0.003 µg/g), and higher for the column fed with HA:TY mixture (0.016 µg/g) and TY (0.024 µg/g), which indicates that labile compounds (tyrosine) highly promotes the removal of Se(IV) during filtration. However, the results revealed that the presence of other metals in the raw water has a negative effect on the adsorption characteristics of Se(IV) during filtration. The adsorption capacity of sand decreased to 0.002, 0.019, and 0.012 µg/g for the columns fed with HA, TY and HA:TY solution, respectively.

Table 4.4. Thomas adsorption parameters of Cu(II) for the columns fed with different NOM types in the presence/absence of metals

	Cu(II) [individual]				Cu(II) [mixed]			
	K_{TH}	q_0	R^2	X^2	K_{TH}	q_0	R^2	X^2
	(L/hr.µg)	(µg/g)			(L/hr.µg)	(µg/g)		
HS	5.00E-05	0.53	0.96	0.80	1.33E-05	1.63	0.95	0.36
TY	2.20E-05	2.89	0.94	0.02	2.53E-05	2.83	0.87	0.14
HS:TY	4.80E-05	1.01	0.96	0.31	2.27E-05	1.90	0.98	0.12
NCTW	3.15E-05	2.01	0.94	0.10	3.87E-05	2.40	0.91	0.22

4.4 DISCUSSION

4.4.1 Removal of HMs during bank filtration

The Pb(II) removal efficiency was higher than that of other metals during filtration, regardless of the prevailing environmental conditions. More specifically, Pb removal efficiency was above 90% under all experimental conditions (different hydraulic rates and feed-water organic compositions). According to Kalakodio et al. (2017), precipitation and adsorption are the main mechanisms which remove Pb(II) during the sand filtration process. Furthermore, a PHREEQC analysis was conducted in this study to determine the precipitation characteristics of the added Pb(II). It was found that its concentration in the feed water was insufficient for precipitation (saturation index SI < 1), which implies that the removal of Pb(II) during the column experiment was mainly caused by adsorption. Sontheimer (1980) reported an average Pb(II) removal efficiency of 75% at BF sites along the Rhine River (Germany). The higher removal efficiency of Pb(II) in this research was mainly attributed to the high temperature of the feed water (30±2 °C), which considerably enhanced the adsorption characteristics of the metal during filtration. Awan et al. (2003) reported that Pb(II) has a high probability to become hydrolysed in water. Thus, it can be readily chemisorbed on the sand. The hydrolysis and adsorption processes were strongly enhanced by the increased temperature. Furthermore, Guanxing et al. (2011) highlighted that the adsorption of Pb(II) onto the soil is an endothermic reaction, and its efficiency is positively related to temperature. The high Pb(II) removal during sand filtration was also observed in a column experiment conducted at 25 °C (Jumean et al., 2010). Thus, it can be concluded that the BF technique is capable of removing Pb(II) from raw water, in particular under hot-climate conditions.

Table 4.5. Thomas adsorption parameters of Se(IV) for the columns fed with different NOM types in the presence/absence of metals

	Se(IV) [individual]				Se(IV) [mixed]			
	K_{TH}	q_0	R^2	X^2	K_{TH}	q_0	R^2	X^2
	(L/hr.µg)	(µg/g)			(L/hr.µg)	(µg/g)		
HS	1.41E-03	0.003	0.91	0.07	2.14E-03	0.002	0.89	0.34
TY	4.64E-04	0.024	0.91	0.03	4.64E-04	0.019	0.87	2.1
HS:TY	5.93E-04	0.016	0.95	0.03	5.29E-04	0.012	0.94	2.2
NCTW	2.14E-03	0.004	0.96	1.99	1.47E-03	0.0017	0.92	0.12

The removal efficiencies of Cu(II), Zn(II), and Ni(II) were lower (65%–95%) than that of Pb(II) during filtration and statistically independent of the flow rate of the feed water. Consequently, the retention time did not play a significant role in the removal of these metals during the filtration process. The efficiency of the BF technique in the removal of these metals under different hydrological and climate conditions has been investigated. Nagy-Kovács et al. (2019) reported removal efficiencies of 59%–99% of these metals at BF sites along the Danube River (Hungary). Awan et al. (2003) highlighted that the removal efficiencies of these metals are highly dependent on their affinity to negatively charge groups (e.g., OH⁻) at the soil surface, which depends strongly on the environmental conditions (e.g. temperature, pH, and organic content of the raw water).

A high persistence of Se(IV) was observed during the column filtration process; the removal efficiency of Se(IV) did not exceed 40% under all experimental conditions. This removal efficiency was statistically independent of the infiltration rate of the feed water. Schmidt et al. (2003) reported a low Se(IV) removal efficiency (15%) at a BF site along the Rhine River (Germany). The lower Se(IV) removal efficiency during the filtration process was mainly ascribed to its natural characteristics. Moreover, Li et al. (2017) reported that Se(IV), which is the predominant species in surface water systems, exhibits a high solubility and low adsorption characteristics. Therefore, its removal efficiency was poor during the filtration process.

4.4.2 Impact of organic matter on HM removal efficiency

The organic composition of the feed water had a minor effect on the Pb(II) removal during filtration. According to Ahmed et al. (2015), organic compounds augment the Pb(II) removal efficiency by forming adsorbable complexes. However, this study revealed that feed water with a higher HA concentration (>15 mg-C/L) might suppress the removal efficiency of Pb(II) by 20%. Likewise, HA had a negative impact on the removal efficiencies of Cu(II), Zn(II), and Ni(II) during filtration. The highest reduction in removal efficiency was observed when the WEOM (higher terrestrial humic content) and DCWW (higher microbial humic content) feed waters were used. It was observed that addition of 20 mg-C/L HA compounds to the feed water reduced the removal efficiencies of Cu(II), Zn(II), and Ni(II) by 40%–75%. Similarly, the Se(IV) removal efficiency decreased with increasing HA concentration in the feed water. The lower removal efficiency of the former metals in the presence of a higher HA concentration was mainly caused by:

(i)   the ability of organic matter to react with metals and to form aqueous complex compounds, which increases the mobility of the metals and reduces their adsorption efficiencies in the sand surface (Zhao et al., 2019);

(ii)   high potential of HA compounds to accumulate on the sand surface, which reduces its adsorption capacity (Refaey et al., 2017b);

(iii)   active functional groups in HA compounds (e.g. carboxylic, phenol, and catechol OH) may associate to the minerals on the sand surface and compete with metal for the adsorption sites (Mal'tseva et al., 2014);

(iv)   the soil structure could be changed by HA compounds by improving its aggregation and thereby reducing the number of adsorption sites (Calace et al., 2009).

Interestingly, the fluorescence data demonstrated that the HA compounds, regardless of their source (microbial or terrestrial), reacted with these HMs and reduced their

85

adsorption efficiency. Moreover, HA compounds enhanced the desorption rates of the HMs via chelating and exchange reaction processes. Tian et al. (2011) illustrated that the desorption rate of Se(IV) increases by approximately 50% in the presence of hydrophobic HA compounds; the hydrophilic compounds could release a low amount (below 3.5%) of the adsorbed Se (El-Said et al., 2011).

The impact of HA compounds on the HMs removal depends strongly on the properties of the metal and its adsorption mechanisms on the sand surface. In this research, Cu(II) was most impacted by the variation in the HA concentration in the feed water. However, Cu(II) exhibited a higher removal efficiency than Zn(II) and Ni(II) in the column fed with low-organic content feed water (NCTW). The sorption of Cu(II) is predominately based on the formation of strong electrostatic and covalent bonds with negative functional groups (e.g. organic matter) on sand surface grains (Refaey et al., 2017a). By contrast, Zn(II) and Ni(II) tend to remain on the surface through an electrostatic attraction process (Refaey et al., 2017b). Therefore, Cu(II) is expected to have a higher affinity than Zn(II) and Ni(II) for the functional groups on the sand grains during the filtration process. However, this affinity decreases with increasing humic concentration in the feed water. Hence, Cu(II) prefers to react with soluble HA substances and to form aqueous complexes than to be adsorbed onto active organic sorption sites of the soil. The Zn(II) and Ni(II) adsorption rates were also affected (to a lower extent) by the humic content of the feed water. Zhao et al. (2019) pointed out that Cu(II) binds stronger to the active groups (e.g. phenols [$-$OH], amines [$-NH_2$], and carboxyl [$-COOH$]) of microbial and terrestrial humic compounds than Zn(II) and Ni(II).

By contrast, the biodegradable (protein-like) compounds enhanced the HM removal efficiency during filtration. According to Abdelrady et al. (2018), LMW organic compounds tend to adsorb faster onto the sand surface than high-molecular weight (HMW) compounds during filtration. Therefore, a metal with a higher affinity towards LMW compounds is more likely to be removed during filtration. Moreover, biodegradable matter promotes biological activity associated with the sand and thus enhances the biosorption and accumulation of metal on its surface (Stefaniak et al., 2018). In this study, the sand column with a higher microbial activity (higher ATP concentration) had a higher HM removal efficiency. Moreover, 5 mg-C/L TY was effective enough to eliminate HMs (Cu, Zn, Ni, and Pb) during the filtration. However, it might enhance the development of an anaerobic environment within the infiltration zone, and consequently, promote the reduction and releasing of HMs into the bank filtrate.

4.4.3 Impact of metals on sorption of Cu and Se during bank filtration

The conducted kinetic experiments revealed that the adsorption rate of Cu(II) increased significantly when the feed water contained a mixture of metals. Hence, the existence of other metals in the feed water retarded presumably the formation rate of the aqueous Cu(II)–HA complex. By contrast, the presence of other metals in the raw water had an adverse effect on the adsorption capacity of the sand for Se(IV) during filtration; a lower Se(IV) removal efficiency was observed, even when the raw water contained LMW (TY) or HMW (humic) organic compounds. Dinh et al. (2017) pointed out that LMW organic compounds tend to react with metals and form stable ring-complexes that compete with Se(IV) for adsorption sites. However, HMW compounds can reduce the adsorption capacity of sand by accumulating and blocking adsorption sites. Therefore, it can be concluded that a better removal of Se(IV) takes place in the infiltration zone when the raw water has lower concentration of metals and higher content of biodegradable organic matter.

4.5 Conclusions

Laboratory-scale column experiments were conducted to assess the impact of the feed water and organic composition on the removal of HMs (Cu, Zn, Ni, Pb, and Se) during the BF process, and the following conclusions are drawn:

- Pb is the most probable metal to be adsorbed into the sand surface during the filtration process. However, humic compounds at concentrations above 15 mg-C/L reduced the adsorption efficiency by approximately 20% in pure water.

- Cu, Zn, and Ni exhibited lower removal efficiency during the filtration process (70%–95%) for the column fed with different water sources. Furthermore, the experimental results showed that 10 mg-C/L humic compounds could reduce the adsorption efficiencies of these metals by 40%–75%.

- Se was the most persistent metal during the column filtration process among the tested metals; its removal was less than 40% under all the experimental conditions.

- Biodegradable matter enhanced the adsorption rates of the aforementioned metals, whereas humic compounds negatively affected their removal efficiencies.

- PARAFAC fluorescence data revealed that presence of both terrestrial and/or microbial humic compounds reduced the adsorption efficiency of the metals.

- The kinetic study revealed that the presence of other metals in the raw water reduced the influence of the humic compounds on the adsorption of Cu and thus enhanced its removal efficiency. Conversely, The presence of HMs exhibited a negative effect on the removal of Se during the filtration process.

In conclusion, this study highlighted the impact of dissolved organic matter composition on the removal efficiencies of HMs, which should be considered in the design and installation of bank filtration wells.

4.6 SUPPLEMETARY DOCUMENTS

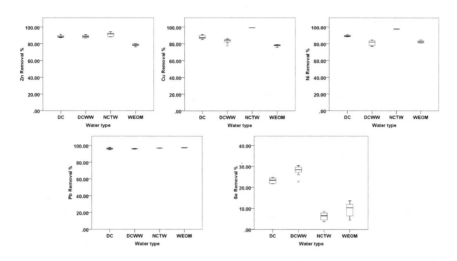

Figure S4.1 .Removal of Cu, Zn, Ni, Pb (initial concentration =150 µg/L) and Se (10 µg/L) during aerobic column infiltration at HLR = 0.6 m/day (temperature=30 °C).

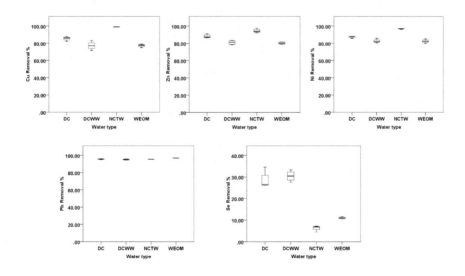

Figure S4.2. Removal of Cu, Zn, Ni, Pb (initial concentration =150 µg/L) and Se (10 µg/L) during aerobic column infiltration at HLR = 0.3 m/day (temperature=30 °C).

Table S4.1. The spectral slopes of the identified PARAFAC fluorescence components and their corresponded components in previous studies from the OpenFluor database (Murphy et al., 2014)

	Ex. Wave. (nm)	Em. Wave. (nm)	Tucker congruence coefficient (TCC)	Previous study	Description
PC1	240,344	474	0.99	(Shutova et al., 2014)	Terrestrial humic (higher molecular weight)
			0.99	(Kulkarni et al., 2018)	
			0.99	(Gonçalves-Araujo et al., 2016)	
			0.99	(Murphy et al., 2011)	
PC2	240,303	396	0.99	(Gonçalves-Araujo et al., 2016)	Terrestrial fulvic /Microbial humic (lower molecular weight)
			0.99	(Kothawala et al., 2014)	
			0.99	(Murphy et al., 2011)	
			0.98	(Stedmon et al., 2007)	
PC3	240, 272	316	0.98	(Gonçalves-Araujo et al., 2016) (Yang et al., 2019)	protein-like (tyrosine and tryptophan -like fluorophores)
			0.98	(Stedmon et al., 2007)	
			0.98	(Wünsch et al., 2018)	

Figure S4.1: Organic fluorescence characteristics of the feed water

Three peaks were identified based on their maximum fluorescence intensity at discrete excitation and emission wavelengths (Baghoth et al., 2011). The spectrum fluorescence matrix of the NCTW spiked with humic displayed two peaks; the first peak (P1, primary humic) was identified at maximum excitation wavelengths between 240-250 nm and maximum emission wavelengths between 380-460 nm. The second humic peak (P2, secondary humic) exhibited maximum fluorescence intensity at excitation and emission wavelengths between 280-320 nm and 400-450 nm, respectively, which implies that the spiked humic material was composed of two fluorescent humic compounds. In the NCTW spiked with tyrosine solution fluorescence matrix one peak (P3) appeared at excitation wavelengths of 250-280 nm and emission wavelengths of 280-350 nm. The NCTW spiked with the mixture (humic and tyrosine) showed three peaks (P1, P2, and P3) at the same wavelengths mentioned above.

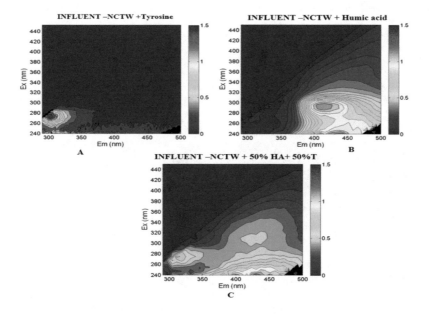

Figure S4.1. F-EEM of: (A) Influent NCTW + TY, (B) Influent NCTW + HA, (C)

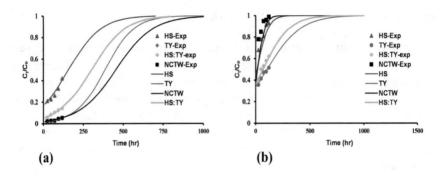

Figure S4.2. Thomas adsorption breakthrough of (a) Cu(II) and (b) Se(IV) for the columns fed with different NOM types

5

REDUCTION OF IRON, MANGANESE AND ARSENIC DURING BANK FILTRATION

ABSTRACT

Bank filtration (BF) has been used for many years as an economical technique for providing high-quality drinking water. However, under anaerobic conditions, the r release of undesirable metal(loid)s such as iron, manganese and arsenic from the aquifer, reduces the bank filtrate quality requiring post-treatment before supply and thus restricts the application of BF. This study investigated the impact of the organic-matter composition of source water on the mobilisation of Fe, Mn, and As during the anaerobic BF process. A laboratory-scale column study was conducted at a controlled-temperature ($30\pm2°C$) using different feed water sourcesat redox conditions between 66 mv and -185 mv. Moreover, batch studies were implemented to study the effect of natural organic matter type (humic, fulvic and tyrosine) and concentration on the mobilisation of the selected metal(loids). The organic matter characteristics of the feed water were elucidated using excitation-emission spectroscopy techniques integrated with parallel factor framework clustering analysis (PFFCA) model.The laboratory experiments demonstrated that the mobilisation of Fe, Mn and As during the BF varied with the organic water concentration and composition of the source water. The fluorescence results revealed that terrestrial and condensed structure humic compounds are more capable to release Fe into the filtrate water. In contrast, Mn exhibited an equal tendency of mobilisation in presence of all the humic compounds regardless of their origin and structure. However, at a humic concentration higher than 5 mg-C/L, Mn showed more affinity towards lower molecular weight humic compounds. Arsenic was found to be the least impacted by the alteration in the source water organic matter composition; its mobilisation was highly correlated with Fe release process. On the other hand, the biodegradable organic matter at high concentration (>10 mg-C/L) was found to be highly effective to turn the infiltration area into Fe-reducing environment and thereby elevating Fe and As concentrations in the pumped water. In conclusion, this study revealed that the DOM composition and concentration of the raw water could play an important role in the mobilisation of metal(loids) during the BF processes.

5.1 INTRODUCTION

Dissolution of the aquifer material and increase in concentration of Fe, Mn, As or F in bank filtrate is well known (Grischek et al., 2017; Poggenburg et al., 2018). Consequently, bank filtrate often need some post-treatment to reduce the concentration of these contaminant below the acceptable value. Redox reactions take place in the infiltration zone is highly determined the quality of bank filtrate. Sequential redox processes occur during infiltration, oxygen is used by microorganisms as the primary electron acceptor for the biodegradation process of organic matter. When the concentration of dissolved oxygen is less than 0.2 mg/L, the microorganisms utilize nitrate as a source of energy followed by manganese, iron, and suplate (Massmann et al., 2008). The reduction and mobilisation of Fe and Mn is considered as major sources of elevated toxic substances (e.g., arsenic (As)) in pumped water, which adversely affects human health and degrades bank filtrate quality (Wang et al., 2012; Yang et al., 2015). The mobilisation processes are mainly controlled by environmental conditions along the subsurface flow path, such as the temperature, hydraulic conductivity of the soil, redox potential, metal speciation in the ores, and the organic and inorganic compositions of source water (Anawar et al., 2003; Kulkarni et al., 2017; Macquarrie et al., 2008; Vega et al., 2017).

DOM is considered as the major energy source for the metal(loid) mobilisation taking place during the subsurface water flow process (Vega et al., 2017). DOM can be subdivided into two major categories namely, biodegradable and non-biodegradable. Biodegradable compounds enrich the biological activity along the flow path and subsequently accelerate the microbial reduction of metal(loid)s into the filtrate water (Schittich et al., 2018). Non-biodegradable compounds (i.e., humics) have high electron shuttle capacities and thus could play a dual role in these mobilisation processes. First, they could enhance the mobilisation of metal(loid)s biotically by mediating the microbial reduction processes (Brune et al., 2004). Second, they might chemically form aqueous complexes with metal(loid)s and thereby increase their concentrations in filtrate water (Liu et al., 2011). On the other hand, the microbial reduction of Fe and Mn might also affect the dynamics and alter the chemical structures of DOM during infiltration. Vega et al. (2017) highlighted that the microbial reduction process is concurrent with the oxidation of labile compounds (e.g., phenol) into refractory humic compounds, as well as the destruction of high-molecular-weight organic compounds into smaller compounds that can serve as electron donors for microorganisms during the microbial reduction processes. The dynamics of DOM and its impact on the mobilisation of metal(loid)s during the subsurface flow of water is highly dependent on the chemical structure of DOM and environmental conditions such as the redox potential and the temperature.

The depletion of oxygen in surface water systems is more likely in arid environments. Hence, there is a higher potential of developing an anaerobic environment during the BF processes, which subsequently leads to the enrichment of pumped water in metal(loid)s and degrades the bank filtrate quality. This is one of the major drawbacks that restrict the widespread application of the BF technique in hot-climate regions (Bartak et al., 2014; Hülshoff et al., 2009; Sprenger et al., 2011). Therefore, this research investigated the role of DOM composition in the mobilisation of Fe, Mn, and As during the BF processes under hot-climate conditions. This is an important step to predict and control their concentrations in pumped bank filtrate water.

5.2 RESEARCH METHODOLOGY

5.2.1 Soil characteristics

Iron coated sand (ICS) used in this research was obtained from the groundwater water treatment plant Brucht (Netherlands). The media was sieved through a 3 mm mesh screen, washed gently with water to discard the deposits, and then dried at 70 °C. The physical and chemical properties of the media were determined and presented in Table 5.1.

Table 5.1. Characteristics of the used media in the column and batch experiments

	Bulk density	Porosity	Organic matter	Fe	Mn	As
Unit	g/cm^3	---	mg-C/g	mg/g	mg/g	mg/g
ICS	1.2±0.1	0.42	10.8±5.4	32.8±1.2	12.1±0.8	0.3±0.1

5.2.2 Column experiment

A laboratory-scale column study was performed to evaluate the role of the organic composition of water in the release of Mn, Fe and As from the soil into the bank filtrate at a high temperature. Four PVC columns with 2.1 cm internal diameter and 50 cm height were developed. The columns were packed at the bottom with a support layer of graded gravel with a diameter 3-5 mm to a height of 5 cm and then filled with ICS (1-3 mm) in deionised water to ensure the homogeneous packing. The feed water was introduced to the columns in an up-flow mode (saturated flow) at a constant hydraulic rate (0.5 m/d) using a variable-speed peristaltic pump. Sub-oxic conditions were maintained in all columns by degassing the feed water tank with a nitrogen stream to dissipate the air until the dissolved oxygen concentration was lower than 0.2 mg/L. Then, the columns were acclimated using Delf canal water (DCW) for more than 70 days. After that, the columns were fed with waters of different organic composition. The first column continued to be fed with DCW. The second column was fed with DCW mixed with secondary treated wastewater effluent from Hoek van Holland, the Netherlands (DCWW). The third column was fed with DCW mixed with water-extractable organic matter (WEOM) that had high humic content. The procedures used for WEOM preparation are described in detail in (Abdelrady et al., 2018). The last column was fed with non-chlorinated tap water (NCTW), representing low-organic-matter-content water.

The experiment was conducted in a controlled-temperature room (at 30 °C). Influent and effluent samples were taken regularly to characterise their quality parameters. Physical and chemical parameters (i.e., pH, redox potential, temperature, and dissolved oxygen concentration) were measured continuously to ensure that the experiment was conducted under the desired environmental conditions.

During this study, four different types of waters were fed to the columns to assess the impact of organic matter composition on the mobilization of Fe, Mn and As during soil passage. The quality parameters of the feed water are summarized in Table 5.2. The redox potential of the influent and effluent water was below zero (ranged between -66 mv and -185 mv) and pH varied between 7.56 and 8.14 during the experimental period.

Table 5.2. Water quality parameters of the influent water

	unit	DC	NCTW	DCWW	WEOM
pH	---	7.73±0.61	7.92±0.54	7.67±0.84	8.1±0.58
DOC	mg-C/L	11.64±3.51	3.98±0.73	11.73±1.06	18.23±1.6
SUVA	L/mg-m	3.52±0.34	1.51±0.18	3.29±0.28	4.85±0.51
NO_3-N	mg-N/L	1.7±0.3	0.2±0.02	1.9±0.2	1.4±0.3
NH_4-N	mg-N/L	0.19±0.04	nd	0.28±0.05	0.86±0.1
PO_4-P	mg-P/L	0.21±0.1	nd	0.25±0.06	0.41±0.12
SO_4^{2-}	mg/L	86±39	42±22	103±29	5.5±0.5
Cl^-	mg/L	74±24	48±18	87±26	12±3

nd: not detected

5.2.3 Batch experiment

Laboratory-scale batch reactors were employed to determine the impact of the type and concentration of natural organic matter (NOM) on the mobilisation of Fe, Mn, and As from the soil into the filtrate water. Each reactor (0.5 L brown glass bottle containing 100 g ICS) was ripened with 400 mL of DCW and the ripening process was continued for 70 days. The feed water was degassed with a nitrogen stream to dissipate the air. Then, the reactors were sealed tightly and placed on a reciprocal shaker at 100 rpm in a controlled-temperature room (30 °C). The feedwater renewal process was conducted inside an anaerobic chamber to ensure that the experiment was conducted under anaerobic conditions. After ripening, three NOM types (humic, fulvic, and tyrosine) were spiked at four different concentrations (5, 10, 15, and 20 mg-C/L) in the reactors (one type and one concentration in each). The humic acid and tyrosine were purchased as powders from Sigma-Aldrich (Netherlands) and the powdered fulvic acid was bought from the International Humic Substances Society. Control (DCW without sand) reactors were operated under the same environmental conditions.

5.2.4 Analytical methods

Influent and effluent samples were collected and analysed directly to avoid the degradation of the organic matter. The samples were first filtered using 0.45 μm filtration (Whatman, Dassel, Germany). The Fe and Mn concentrations were determined using inductively coupled plasma–mass spectrometry instrument (ICP-MS) (Xseries II Thermo Scientific, Bermen, Germany). The arsenic concentration was analysed using a graphite furnace atomic absorption spectrophotometer (Solaar MQZe, Thermo Electron Co). The limit of quantitation (LOQ) of Fe, Mn and As was 10, 10 and 2 μg/L, respectively.

The organic matter content of the feed waters was measured as DOC (in mg-C/L) using the combustion technique with a total organic carbon analyser (TOC-VCPN (TN), Shimadzu, Japan). The organic composition of the influent and effluent water was determined using the fluorescence excitation–emission technique (EEM). A Fluoromax-3 spectrofluorometer (HORIBA Jobin Yvon, Edison, NJ, USA) was used to determine the fluorescence intensity (FI) of the water samples at excitation wavelengths (λex) of 240–452 nm (interval = 4 nm) and emission wavelengths (λem) of 290–500 nm (interval = 2 nm). Fluorescence indices, including the humification index (HIX), the fluorescence index (FIX), and the biological index (BIX), were estimated based on the EEM measurements and used to have insight information about the organic characteristics of the feedwaters.

5.2.5 Fluorescence modelling

Although fluorescence-EEM is a qualitative technique, it is still effective in quantifying the characteristics of DOM. A parallel factor framework-clustering analysis (PFFCA) model was recently stipulated by Qian et al. (2017) for decomposing the EEM dataset into major fluorescence components representing different NOM compounds. Briefly, PFFCA decomposes the EEM datasets into several factors (7–13). Afterwards, the highly correlated factors are clustered into one component representing organic matter with the same fluorescence characteristics.

To ensure the quality of the data and the robustness of the fluorescence dataset interpretation, parallel factor (PARAFAC) analysis was used as well to decompose the fluorescence dataset into major components and to assess the contribution of each component to the full fluorescence spectrum. The development and validation of the model developing were described in details in Murphy et al. (2013).

The redox index (RI) is used as an indicator of the reducibility of quinone-like moieties in the DOM of raw water. This parameter was calculated following the definition of Miller et al. (2006), by dividing the sum of the FI of the reduced quinone-like moieties to the total of the (reduced and oxidised) quinone-like moieties of the fluorescence components, with values close to 1 indicating a higher reducing capacity of the feed water (Gabor et al., 2014).

5.2.6 Statistical analysis

The relationship between Fe, Mn, and As concentrations in the effluent waters and the fluorescence organic characteristics of the feed water were explored using nonparametric Spearman rank correlation (ρ) analysis. The strength of the correlation was considered high when the correlation value (ρ) was higher than 0.7 and moderate when it ranged between 0.7 and 0.5 (Hinkle et al., 1988). The relationship was considered significant if the criterion level of significance (p) was lower than 0.05.

5.3 RESULTS

5.3.1 Impact of DOM composition on metal(loids) mobilization (column experiment)

Fluorescence components

A fluorescence EEM dataset of 80 water samples collected from the influents and effluents of the columns was used to develop and validate a PFFCA-EEM model. The dataset was decomposed into 11 factors and afterwards clustered into four main components (FC1–FC4). To ensure the reliability and interpretability of the fluorescence EEM dataset, it was also analysed using the PARAFAC technique. A model with four fluorescence components (PC1–PC4) was successfully split-half-validated and explained 99.7% of the data variability. The footprints of the PFFCA components and their corresponding PARAFAC components is presented in Figure 5.1 and Figure S5.1. A comparison between the spectral characteristics of the recognised components and those reported in earlier studies was conducted using OpenFluor online database (Table S5.1). Component FC1 exhibited maximum wavelengths of λex = 332 nm and λem = 460 nm and therefore it could be allocated to a terrestrial humic component (Coble, 1996). Singh et al. (2013) reported that this component is ubiquitous in a reduced environment and holds reduced quinone-like moieties in its structure. Component FC2 showed a maximum peak at λex = 310 nm and λem = 410 nm, and therefore it was as attributed to a combination of terrestrial fulvic acid and microbial humic compounds (Gonçalves-Araujo et al., 2016; Li et al., 2016). Component FC4 displayed a maximum peak at λex = 300 nm and λem = 410 nm. This component is frequently associated with marine and/or microbial humic fluorophores and was recently connected to humic/fulvic compounds of agricultural origin (Baghoth et al., 2011). According to Singh et al. (2013), component

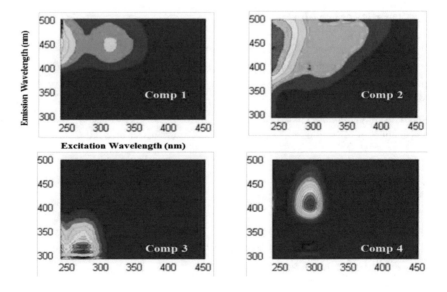

Figure 5.1. Contour plots of the four PFFCA components (PF1-PF4) identified from the complete measured F-EEMs dataset for the influents and effluents of column experiment.

FC2 and component FC4 contain oxidised moieties in their structures. Based on these definitions, the RI was calculated as the ratio of the first component (FC1) to the sum of the three humic components (FC1, FC2, and FC4). Component FC3 exhibited maximum fluorescence at the shortest wavelengths (λex = 240 nm, 268 nm) and (λem = 308 nm), which corresponds to a protein-like fluorophore (tyrosine and tryptophan-like fluorophores) (Wünsch et al., 2017).

Feed water DOM characteristics

Four feed waters (DCW, DCWW, WEOM, and NCTW) with different organic compositions were used in the experiments. Table 5.3 shows that WEOM had the highest concentration of organic matter (DOC ranged between 15.8 and 19.3 mg-C/L) and the NCTW influent had the lowest one (DOC varied from 1.98 to 4.1 mg/L). However, the DOC levels of DCW and DCWW were 10.8–13.6 mg-C/L and 10.2–13.7 mg/L, respectively.

The fluorescence indices were used to elucidate the DOM characteristics of the feed waters (Table 5.3). The results showed that DCW and DCWW had relatively similar fluorescence DOM characteristics. The HIX, FIX, and BIX of the DCW feed water were 0.85, 1.24, and 0.54, respectively, whereas they were 0.83, 1.26, and 0.61 for DCWW. WEOM had the highest humic content (HIX = 0.9) and the lowest microbial-derived (autochthonous) DOM content (BIX = 0.47), and its ratio of terrestrial humic DOM to microbial DOM (FIX) was 1.08. By contrast, NCTW retained a lower humic content (HIX = 0.62) and higher fresh DOM (BIX = 0.73). There were no notable changes in the fluorescence indexes values for DCW, DCWW, and WEOM feed water during the filtration process. However, an increase in the terrestrial humic content (HIX = 0.79, FIX = 1.2) for the NCTW effluent was recorded.

The PFFCA-EEM technique was used to gain insight into the DOM characteristics of the influent waters. Table 5.3 shows the average FI of PFFCA-EEM components for all of the feed-water types. The results demonstrated that the WEOM influent had a significantly higher concentration of humic compounds than the other influents. Terrestrial humic (FC1) was present in a proportion (45–49%) of the fluorescence spectrum of WEOM feed water; however, it only represents 43–46%, 41–42%, and 33–34% of the DCW, DCWW, and NCTW fluorescence spectrum, respectively. The average FIs of microbial humic FC2 and humic-like from agricultural sources, FC4, were 2.36±0.42 and 0.24±0.04 RU, while their values for NCTW were 0.2±0.09 and 0.1±0.03 RU, respectively. The humic-like components (FC1, FC2, and FC4) exhibited a decreasing behaviour during the filtration; the FI of the humic components decreased by 8–26% for DCW, 11–24% for DCWW, and 20–32% for WEOM influent water. An exception was the terrestrial and processed humic compounds of the NCTW influent, which followed an increasing behaviour during the filtration process. On average, the FI of FC1 and FC2 increased by a factor of two and three, respectively, during the filtration process.

Table 5.3. The organic matter characteristics of the feed waters

Parameter	Units	DCW	NCTW	DCWW	WEOM
DOC	mg-C/L	11.64±3.51	3.98±0.73	11.73±1.06	18.23±1.6
HIX	----	0.85±0.02	0.64±0.04	0.83±0.06	0.9±0.02
FIX	----	1.25±0.07	1.35±0.02	1.26±0.14	1.08±0.13
BIX	----	0.57±0.11	0.73±0.04	0.61±0.17	0.47±0.11
FC1	RU	1.48±0.22	1.44±0.24	0.23±0.09	2.63±0.64
FC2	RU	1.10±0.19	1.53±0.21	0.20±0.09	2.36±0.42
FC3	RU	0.47±0.19	0.52±0.14	0.15±0.08	0.65±0.23
FC4	RU	0.20±0.03	0.15±0.06	0.10±0.03	0.24±0.04

DCWW and WEOM had the highest concentration of protein-like fluorescent compounds. The average FI of FC3 was 0.47±0.19, 0.52±0.14, 0.15±0.08, and 0.65±0.23 RU for DCW, DCWW, NCTW, and WEOM influents, respectively. For all feed-water types, the protein-like component FC3 exhibited a decreasing trend during the filtration process; its removal ranged between 67% and 80%, which is in agreement with Abdelrady et al. (2019), who observed an average removal of 80% of labile organic compounds during the filtration process.

Iron, manganese and arsenic mobilisation

Figure 5.2 shows the concentrations of Fe, Mn, and As in the effluents of the columns fed with different water sources. It shows that the mobilisation of Mn was much higher than that of Fe and As. The concentration of Fe, Mn, and As in the effluent for the four different water types were 10–20 µg/L, 1500–3900 µg/L, and >2–7.1 µg/L, respectively. The concentrations of the tested metal(loid)s in the effluents were found to be dependent on the feed-water source. The WEOM feed water exhibited the highest capacity to release Fe, Mn, and As from The ICS into the filtrate water. The mean concentration of Fe, Mn, and As in the WEOM effluent were 18.28±1.5 µg/L, 3590±185.3 µg/L, and 5.48±1.2 µg/L, respectively. The DCW and DCWW effluents showed a lower capacity of releasing these metal(loid)s during the filtration process. The Fe, Mn, and As concentrations of the DCW effluent were 11.9–15.1 µg/L, 3100–3600 µg/L, and 3.7–5.1 µg/L, respectively, and those of the DCWW effluent were 12.1–13.9 µg/L, 3000–3600 µg/L, and 2.9–4.7 µg/L, correspondingly. By contrast, NCTW exhibited the lowest capability of releasing the assessed metal(loid)s, with Fe and Mn concentrations of 10.98±0.9 µg/L and 1990±384.3 µg/L, respectively. Moreover, the As concentration was always below the detection limit (<2 µg/L). A strong correlation ($\rho = 0.77$) was observed between Fe and As concentrations in the effluent waters. By contrast, a weak correlation ($\rho = 0.35$) was detected between the As and Mn concentrations.

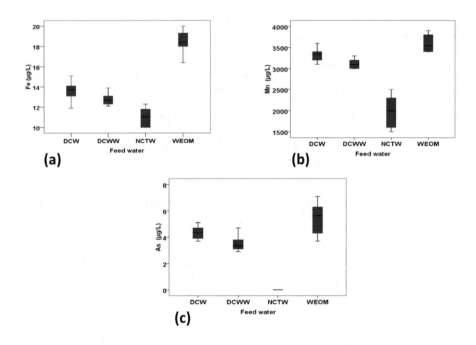

Figure 5.2. Mn, Fe and As concentrations in the effluents of columns operated using different water sources (DCW, DCWW, WEOM, NCTW) (anaerobic, 30°C, HLR=0.5 m/d)

103

5.3.2 Relationship between DOM composition and metal(loid) mobilisation

Figure 5.3 illustrates that the release of Fe, Mn, and As during the filtration is highly dependent on the DOM concentration of the feed water. The feed DOM concentration was highly correlated with Fe (ρ=0.89) and Mn (ρ=0.78) concentrations in the effluent

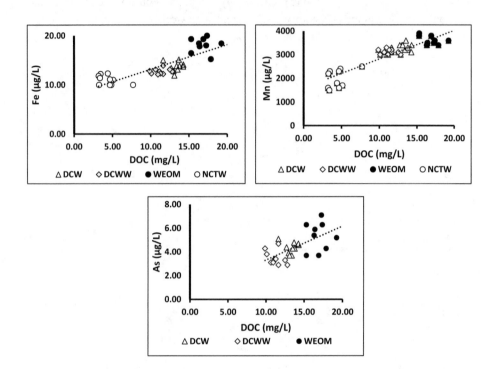

Figure 5.3. Impact of DOC concentration (mg-C/L) of different water sources (DCW, DCWW, WEOM, NCTW) on Fe, Mn and As mobilization during column

waters, whereas a moderate positive relationship was observed between DOM concentration and effluent As concentrations ($\rho = 0.63$).

The fluorescence indexes results illustrate that Fe, Mn, and As release during the anaerobic filtration is much more dependent on the humic content of the feed water than its autochthonous microbial DOM concentration (Figure 5.4). The correlation between HIX and the effluent Fe concentration was observed to be $\rho = 0.87$, whereas it was recorded as $\rho = 0.75$ for the Mn concentration. However, a moderate correlation was observed between HIX and the effluent As concentration ($\rho = 0.51$). On the other hand, the mobilisation of Fe, Mn, and As was observed to be in a negative relationship with the microbial and terrestrial DOM ratio, and the correlation between the FIX of the feed waters and the released metal(loid)s was $\rho = -0.71$ for Fe, $\rho = -0.64$ for Mn, and $\rho = -0.39$ for As.

The correlation between the released metal(loid) concentrations and the FI of the fluorescence components of the feed waters was estimated and is presented in Figure 5.4 and figures S5.2-S5.4. A strong positive correlation was observed between the Fe concentration and the FI of the humic components (FC1, FC2, and FC4) in the feed waters. The Mn mobilisation, as well, exhibited a strong correlation with the FI of humic components from different sources; the correlations between the Mn concentration and the FI of the PFFCA humic components (FC1, FC2, and FC4) were 0.78, 0.77, and 0.76, respectively. Conversely, the As concentration in the column effluent water was found to be in a moderate correlation with terrestrial humic components and in weak correlations with the other humic components (FC2 and FC4). Furthermore, the RI demonstrated a moderate correlation with Fe mobilisation and a weak correlation with Mn and As mobilisation. This infers a role for condensed-structure humic compounds in the mobilisation of Fe into filtrate water. On the other hand, labile fluorescence compounds (FC3) were found to have a greater influence on the mobilisation of Mn than on those of Fe and As.

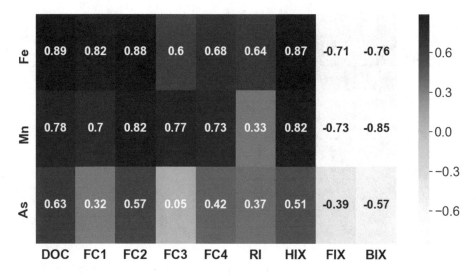

Figure 5.4. Correlations between the tested metal(loid)s and the fluorescence characteristics of the feed waters

5.3.3 Impact of NOM concentrations on metal(loid) mobilisation (batch study)

The effect of NOM (humic, fulvic, and tyrosine) concentrations on the release of Fe, Mn, and As from the soil into the filtrate water under anaerobic conditions was studied in established batch reactors. The experimental results illustrated that the Fe mobilisation increased steadily with humic and fulvic concentrations (Figure 5.5a). Fe exhibited relatively higher affinity with humic (high molecular weight) compounds compared to that with fulvic (low molecular weight) compounds. The concentration of soluble Fe reached 162±19 µg/L when the humic concentration was 20 mg-C/L, whereas it was 114±28 µg/L at the same concentration of fulvic acid. On the other hand, labile compounds (i.e., tyrosine) exhibited lower ability to release Fe at low concentrations; the concentrations of Fe in the effluent water in the batch reactors injected with 5 and 10 mg-C/L of tyrosine were 30±17 and 32±11 µg/L, respectively. However, an immense increase in Fe concentration was observed for the batch reactors injected with 15 mg-C/L and 20

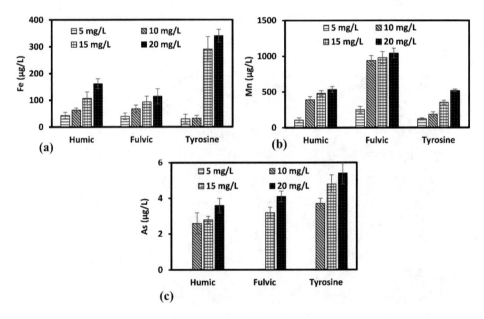

Figure 5.5. Effect of NOM type (humic, fulvic, and tyrosine) and concentration (5, 10, 15, 20 mg-C/L) on (a) Fe, (b) Mn, and (c) As mobilisation during the batch experiment (anaerobic, 30 °C)

mg-C/L of tyrosine, where the soluble Fe concentrations were 290±47 µg/L and 340±24 µg/L, respectively.

Figure 5.5 b illustrates the impact of NOM type and concentration on the mobilisation of Mn during the filtration process. It can be observed that fulvic compounds have the highest capacity of releasing Mn into the filtrate water. The concentration of Mn increased from 250 µg/L to 1041 µg/L as the concentration of fulvic acid increased from 5 mg-C/L to 20 mg-C/L. The same trend was noted for humic compounds, where a humic acid concentration of 20 mg-C/L resulted in increasing the released Mn concentration five times compared with its effluent concentration in the reactors injected with 5 mg-C/L humic acid. Likewise, a positive relationship was observed between Mn concentration in the batch reactor effluents and tyrosine concentration. The concentration of Mn was increased by 35%, 65%, and 77% in the batch reactors injected with 10, 15, and 20 mg-C/L, respectively, compared with its concentration in the batch reactors injected with 5 mg-C/L.

The batch study results showed that the mobilisation of As is obviously affected by the type and concentration of NOM (Figure 5.5c). NOM at low concentration exhibited a poor capacity of releasing As from its precipitated form into the filtrate water. No As concentration was quantitatively (LOQ = 2 µg/L) detected in any of the batch reactor effluents at DOC = 5 mg-C/L for the three NOM types. At higher concentrations (20 mg-C/L), tyrosine exhibited the highest capacity of releasing As into the filtrate water; the As concentrations in the batch reactors injected with humic, fulvic, and tyrosine were 3.6±0.4, 4.1±0.3, and 5.4±0.6 µg/L, respectively after the residence time of 30 days.

5.4 DISCUSSION

5.4.1 Mobilisation of Fe, Mn and As during BF

The mobilisation of Mn was two orders of magnitude higher than that of Fe during the infiltration process under the experimental conditions. From stoichiometry, 1 mole of simple organic matter (e.g., acetate) could act as an electron donor to mobilise 4 moles of Mn and 8 moles of Fe from the soil into the filtrate water (Lovley et al., 1988). However, the higher mobilisation of Mn compared with those of Fe and As observed during this study was probably due the combination of three factors:

(i) The reducing environment was still not high enough to be dominated by the Fe microbial reduction process.

(ii) Mn oxide was used as an agent for the Fe oxidation process and thereby could be precipitated as Fe oxides. This process would increase the re-adsorption of As on the surface of iron oxides and thus reduce the concentrations of Fe and As in the filtrate water (Vega et al., 2017). Neidhardt et al. (2014) illustrated that Fe (oxyhydr)oxide redox processes control the concentration of As in filtrate water.

(iii) The column experiment was conducted at a high temperature (30 °C) and a relativley low infiltration rate (0.5 m/d), which greatly enhanced the microbial reduction of manganese.

Low infiltration rate increases the contact time and interaction between the metals of the soil and the DOM of the raw water. Moreover, it promotes the reducing conditions along the flow path, which leads to increase the mobilisation of metals into the infiltrating water. Paufler et al. (2018) conducted column studies to assess the impact of the temperature (15–35 °C) and flow rate on the Mn mobilisation during the BF process. It was found that raising the temperature from 20 to 30 °C at a low hydraulic rate (1 ml/min) resulted in increasing the Mn release rate (K_{Mn}) 10–15 times, whereas K_{Mn} was negligible when the feed-water temperature was below 10 °C. Bourg et al. (1994) reported that increasing the temperature (>10 °C) triggers the microbial reduction of manganese in an alluvial aquifer. Based on these findings, it is presumed that the higher Mn mobilisation rate in the column experiment relative to Fe and As mobilisation is primarily ascribable to the microbial reduction of Mn at high temperature (30 °C) that stimulates simultaneously the associated Fe (II) and As (III) oxidation and precipitation processes. This plausible mechanism was observed and reported in many BF and alluvial aquifer fields (Bourg et al., 1993; Bourg et al., 1994; Eckert et al., 2006; Kedziorek et al., 2008; Vega et al., 2017) . However, it is worth noting that other mechanisms (e.g., the simultaneous mobilisation of Fe and Mn) have also been observed in many BF fields (Grischek et al., 2017; Hamdan et al., 2013; Matsunaga et al., 1993) . It should also be recognized that altering the surrounding environmental conditions (e.g., pH, alkalinity, infiltration water redox, raw water chemistry and soil minerals types) can change the mobilization behaviours of the metals being studied.

5.4.2 Influence of composition and concentration of DOM on metal(loid) mobilisation

The strong correlation between DOC and metal(loid) (Fe, Mn, and As) concentrations in the effluent water of the column indicates that DOM plays a key role in the mobilisation of these metal(loid)s during infiltration. In this research, NCTW with low organic content demonstrated lower efficiency in releasing Fe, Mn, and As into the filtrate water compared with those of other source-water types. This is consistent with the findings of recent studies (Hossain Md Anawar et al., 2013; Vega et al., 2017), which reported an increase in the mobilisation rate of Fe and As associated with higher DOM in Daudkandi (Meghna delta) and Marua of the Ganges delta plain aquifer in Bangladesh (Hossain M. Anawar et al., 2003; Hossain Md Anawar et al., 2013). Wang et al. (2012) illustrated that DOM influences the mobilisation and release of metal(loid)s during the passage of the soil through a redox process, complexation, and competitive adsorption. Therefore, it can be concluded that the organic composition of feed water controls the concentration of Fe, Mn, and As in bank filtrate water.

Refractory compounds (i.e., humic) play a vital role in the mobilisation of Fe, Mn, and As during the subsurface flow of water. The experimental results revealed that the mobilisation of Fe and Mn has a strong positive relationship with the humic content of the feed water (HIX), whereas As mobilisation was the least affected by changes in the humic content of the feed water. Several previous studies highlighted the multiple roles of humic compounds in the mobilisation of metal(loid)s during the filtration process (Brune et al., 2004; Liu et al., 2011; Yuan et al., 2018). Owing to their high electron shuttle capacities, humic compounds might act as catalysts for the iron microbial reduction process by transferring electrons between the insoluble metal(loid)s and the reducing microorganisms; this was investigated in many field studies (Mladenov et al., 2010; Mladenov et al., 2015; Poggenburg et al., 2018). In addition, these electron-rich compounds have high capabilities of binding to the soil metal(loid)s and increasing their solubility (Sharma et al., 2010). Chen et al. (2003) pointed out that humic compounds are more soluble at higher pH (>4); thus, they are subjected to more trapping of metal(loid)s from the soil and are maintained in their soluble forms by forming soluble complex compounds.

The effect of humic compounds on the metal(loid) mobilisation is relatively dependent on the origin and characteristics of the organic compounds and type of metal(loid). In this research, a moderate correlation was detected between RI and Fe concentration in the filtrate water, indicating that terrestrial humic compounds (with condensed structures) have relatively higher capability to trap Fe into filtrate water than lower molecular weight humic compounds. According to Yuan et al. (2018), high-molecular-weight humic compounds have more binding sites, and therefore higher capability of forming chemically stable compounds with Fe, than less-condensed-structure and labile compounds. Therefore, it can be deduced that humic compounds of terrestrial origin are more triggered to release iron from its solid forms into filtrate water. This was confirmed by the batch experiment results, which showed a relatively higher increase in Fe mobilisation with humic than with fulvic acid (lower-molecular-weight humic compounds). By contrast, a weak correlation was observed between RI and Mn mobilisation, demonstrating comparable abilities of humic compounds with low and high condensation (i.e., terrestrial and microbial humic) to mobilise Mn; this indicates that Mn metal has a high affinity to both low- and high-molecular-weight humic compounds. This is in agreement with Vega et al. (2017), who observed higher concentrations of Mn to be associated with higher humic content in shallow aquifers at West Bengal (India). Nevertheless, the batch experiment showed that fulvic compounds at higher concentrations (>5 mg-C/L) have a higher capability of mobilising Mn than higher-molecular-weight humic compounds; this could be due to their high capacity to form soluble and stable Mn complexes at high pH (>6) compared with humic compounds, as proven previously by Du Laing (2010).

The mobilisation of Mn during the filtration process was observed to be highly influenced by the biodegradable organic matter content of the source water. The high correlation between Mn concentration in the filtrate water and protein-like fluorescent components indicates that the microbial reduction of Mn is the primary mechanism of Mn release. However, a moderate correlation was observed between Fe concentration and protein-like components. The batch experiment showed that a protein-like compound (i.e., tyrosine) is the most effective NOM for releasing Fe and As at high concentrations (>10 mg-C/L). This is probably due to the increased concentration of biodegradable compounds, which increases the biological activity associated to sand, increases the mobilisation of Mn and decreases the redox potential to a level that permits bacteria to use Fe from the soil as a source of energy, thus increasing the Fe concentration in the effluent water. This process is concurrent with the release of As adsorbed on Fe (oxyhydr)oxides into the filtrate water.

5.5 CONCLUSIONS

The mobilisation of geogenic metal(loid)s, such as Fe, Mn, and As, during the BF process restricts their application, particularly in arid-climate countries. Based on the results of laboratory-scale column and batch studies conducted to assess the impact of the organic composition of raw water on the mobilisation of these metal(loids) during anaerobic BF, the following conclusions are drawn:

- The DOM concentration and composition of the raw water could change the capacity of mobilisation of metal(loids) during the filtration process

- The mobilisation of Mn under the applied experimental conditions was two and three orders of magnitude higher than those of Fe and As, respectively.

- The humic content of the source water was found to significantly affect the release of Fe and Mn during filtration; a positive relationship was found between the mobilisation of the metal(loids) and the HIX of the source water.

- Terrestrial humic compounds with complex structures showed a higher capability of releasing Fe from the soil into the filtrate water. On the other hand, source water with lower-molecular-weight humic compounds at a concentration of >5 mg-C/L was able to mobilise Mn at a higher rate than higher-molecular-weight humic compounds. However, the metal(loid) species in the soil may alter the behaviours of NOM compounds, which should be investigated.

- Arsenic mobilisation was observed to have a high (positive) correlation with Fe mobilisation and a weak correlation with the variations in the organic matter composition of source water.

- Biodegradable organic matter was found to be effective in mobilising Mn into the filtrate water; a strong correlation was observed between the FI of protein-like components in the source water and Mn mobilisation. However, a moderate correlation was found between Fe mobilisation and the FI of protein-like components. Nevertheless, the experimental results showed that a high concentration (>10 mg-C/L) of a protein-like compound is sufficient to produce a Fe-reducing environment in the infiltration area and thereby increase the Fe concentration in the filtrate water.

- In summary, this study revealed that the DOM composition of source water determines the redox environment during the BF process and affects the mobilisation process of the metal(loid)s that should be considered during the BF design process. Moreover, this research highlights the efficiency of the fluorescence spectroscopy technique as a monitoring tool for characterisation the DOM of the raw water as well as for prediction and control the redox process and the mobilisation of metal(loid)s during BF.

5.6 SUPPLEMENTARY DOCUMENTS

Annex A: Comparsion between PARAFAC and PFFCA models

PARAFAC is a multi-way statistical technique that uses an alternating least squares algorithm to decompose the fluorescence dataset into trilinear terms and a residual array as described by Andersen et al. (2003). Each term represents a group of fluorescent organic compounds. Models with (3–7) fluorescence components were tested and different validation tools (i.e., split-half validation and residual error) were used to define the right number of components. PFFCA-EEM and PARAFAC-EEM models were developed and validated using the N-Way and drEEM MATLAB toolboxes in MATLAB (version 8.3, R2014a).

It can be observed (Figure S5.1) that there is no remarkable dissimilarity between the two protein-like components (FC3 & PC3). Likewise, microbial humic components (FC4 & PC4) showed similar spectral characteristics. By contrast, PFFCA terrestrial humic component FC1 peak exhibited a blue-shifted emission spectrum compared to the spectral characteristics of the corresponded PARAFAC component PC1, implying that FC1 encompass organic compounds with less condensed structure than PC1. The maxima λex/λem wavelengths for PC1 component was 332/480 nm. On the other hand, the other PFFCA humic component FC2 covered a wider range of excitation and emission wavelengths than its corresponded PARAFAC component PC2, referring that it is a mixture of two or more humic fluorophores. PC2 maxima λex and λem were taken place at 308 and 420 nm, respectively.

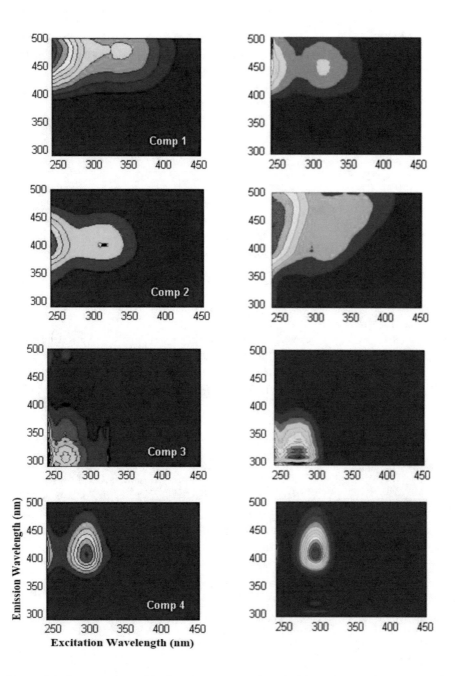

Figure 5.1. Contour plots of the four components identified from the complete measured F-EEMs dataset for the influents and effluents water of column experiment.

Table S5.1. The spectral slopes of the identified PARAFAC fluorescence components and their corresponded components in previous studies from the OpenFluor database (Tucker congruence coefficient (TCC=0.99) (Murphy et al., 2014)

	Ex. Wave. (nm)	Em. Wave. (nm)	Previous studies	Traditional classification (Coble, 1996)	Description
PC1	240, 340	480	(Shutova et al., 2014), (Osburn et al., 2016a), (Yamashita et al., 2010), (Murphy et al., 2011),	Peak A	Terrestrial humic (higher molecular weight)
PC2	240, 308	402	(Gonçalves-Araujo et al., 2016), (Shutova et al., 2014), (Murphy et al., 2011), (Li et al., 2016)	Peak M + Peak A	Terrestrial fulvic /Microbial humic (lower molecular weight)
PC3	240, 268	308	(Osburn et al., 2016a),(Wünsch et al., 2017), (Gonçalves-Araujo et al., 2016)	Peak T+ Peak B	protein-like (tyrosine and tryptophan -like fluorophores)
PC4	240, 296	408	(Walker et al., 2009), (Li et al., 2016), (Kowalczuk et al., 2013)	Peak M	Marine/microbial humic

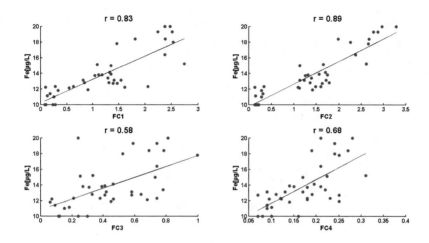

Figure S5.2. Correlation between the FI of the PFFCA components and the Fe concentration of the effluent water (column study, 30 °C, anaerobic, HRL = 0.5 m/d)

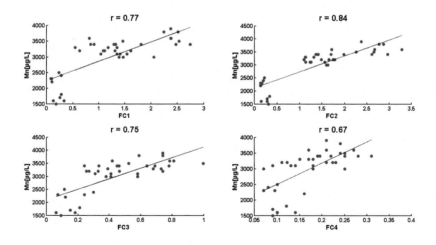

Figure S5.3. Correlation between the FI of the PFFCA components and the Mn concentration of the effluent water (column study, 30 °C, anaerobic, HRL = 0.5 m/d)

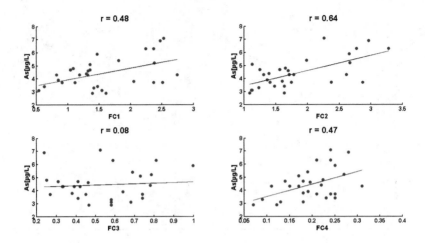

Figure S5.4. Correlation between the FI of the PFFCA components and the As concentration of the effluent water (column study, 30 °C, anaerobic, HRL = 0.5 m/d)

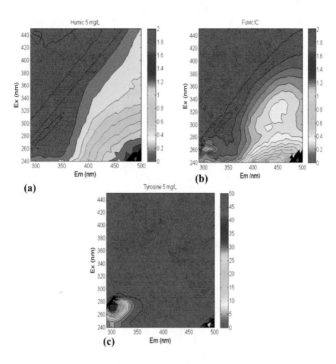

Figure S5.5. F-EEM spectra analysed for humic acid (a), fulvic acid (b) and tyrosine, initial concentration (5 mg-C/L)

6

ANALYSIS OF THE PERFORMANCE OF BANK FILTRATION FOR WATER SUPPLY IN ARID CLIMATES: CASE STUDY IN EGYPT

ABSTRACT

Bank filtration (BF) is acknowledged as a sustainable and effective technique to provide drinking water of adequate quality; it has been known for a long time in Europe. However, this technique is site-specific and therefore its application in developing countries with different hydrologic and environment conditions remains limited. In this research, a 3-discipline study was performed to evaluate the feasibility of the application of this technique in Aswan City (Egypt). Firstly, a hydrological model was developed to identify key environmental factors that influence the effectiveness of BF, and to formulate plans for the design and management of the BF system. Secondly, water samples were collected for one year (January 2017 to December 2017) from the water sources and monitoring wells to characterize the bank-filtrate quality. Lastly, an economic study was conducted to compare the capital and operating costs of BF and the existing treatment techniques. The results demonstrated that there is high potential for application of BF under such hydrological and environmental conditions. However, there are some aspects that could restrict the BF efficacy and must therefore be considered during the design process. These include the following: (i) Over-pumping practices can reduce travel time, and thus decrease the efficiency of treatment; (ii) Locating the wells near the surface water systems (<50 m) decreases the travel time to the limit (<10 days), and thus could restrict the treatment capacity. In such case, a low pumping rate must be applied; (iii) the consequences of lowering the surface water level can be regulated through the continuous operation of the wells. Furthermore, laboratory analysis indicated that BF is capable of producing high quality drinking water. However, an increase in organic matter (i.e., humics) concentration was observed in the pumped water, which increases the risk of trihalomethanes being produced if post-chlorination is implemented. The economic study ultimately demonstrated that BF is an economic and sustainable technique for implementation in Aswan City to address the demand for potable water.

6.1 INTRODUCTION

Access to potable drinking water is a major challenge confronting water service providers in arid and semi-arid countries owing to dwindling water quantity and quality. This issue is more pressing in Egypt, where the annual growth rate of the population is high (1.9%) and the surface water systems are extremely polluted (Aboulroos et al., 2017). 128 agricultural and industrial drains discharge water with a high load of chemical pollutants to the Nile River (NR) (Wahaab et al., 2019). The concentration of organic matter ranges between 2.3 and 11.3 mg/L in the Nile Delta (Badr, 2016). Hence, there is a high potential for the formation of carcinogenic trihalomethane compounds during the treatment process. Therefore, the Egyptian government has recently relied on bank filtration (BF) as a robust and economical technique to replace or integrate with existing waterworks to provide drinking water of adequate quality (Wahaab et al., 2019).

BF is an affordable natural treatment technique, where river water naturally flows through the riverbanks to an aquifer. A sequence of chemical, physical, and biological processes occurs during the sub-surface flow and reduces pollutant concentrations. It is a water treatment process that is environmentally sound and attenuates contaminant concentrations without chemicals addition (Ray et al., 2002). This simple technique has proven its effectiveness to improve water quality and increase its biological stability. Sandhu et al. (2019) reported a 50% removal of dissolved organic matter (DOM) and 13–99% of the micropollutants at BF fields along the Yamuna River (India). The BF principle has evolved in Europe and has been extensively used over hundred years along the Elbe and Rhine rivers for domestic water production. Recently, the application of this technique has been extended for countries such as India (Boving et al., 2018), China (Pan et al., 2018), Brazil (Romero-Esquivel et al., 2017) and Egypt (Wahaab et al., 2019) with different hydrological and environmental conditions. However, bank filtrates contribute less than 0.1% to Egypt's national water supply network (Ghodeif et al., 2016). Thus, there is a requirement to assess the performance of this technique under these local hydrological conditions to propose guidelines to promote the application of this sustainable technique.

The effectiveness of BF is highly influenced by the hydrological and hydrogeological conditions of the surface water system and aquifer. Different studies (Bartak et al., 2014; Ghodeif et al., 2016; Hamdan et al., 2013; Shamrukh et al., 2011) demonstrated the hydrological conditions at NR are favourable for the application of the BF technique. Such conditions were defined by Wahaab et al. (2019) as follows: (i) The aquifer has high hydraulic conductivity ($K > 1 \times 10^{-4}$ m/s) and appropriate thickness (>10 m), reflecting a high water transmission capacity. (ii) The flow of the NR is erosive, which reduces the riverbed's clogging development, and thus, (iii) the NR is well connected with the adjacent aquifer. However, the interaction between the NR and the aquifer is projected to change shortly due to the anthropogenic activities (e.g. construction of the Renaissance

121

dam in Grand Ethiopia (GRED) and the impacts of climate change which could reduce the availability of surface water and consequently impact the performance of BF. Recent studies (Abdelhalim et al., 2020; El-Nashar et al., 2018) have demonstrated that the construction of GRED will reduce the storage capacity of the Aswan High Dam (AHD) by 14.8–60.7%. In comparison, as a consequence of the decrease in the discharge of the AHD by 10%, the NR water level is estimated to decline by 0.45–0.75 m. Regarding climate change, Beyene et al. (2010) reported that precipitation in the Nile Basin is projected a decrease up to 40% by the end of the 21st century. Hence, the annual inflow to the AHD Lake is expected to decrease by 16%. The decline in the availability of surface water could influence the BF performance parameters, such as travel time and drawdown, and reduce the effectiveness of treatment process.

The quality of the bank filtrates is highly dependent on the quality characteristics of the source waters, environmental conditions of the infiltration zone, and the design parameters of the BF system (i.e., the distance between the wells and river, number of wells, and production capacity). Wahaab et al. (2019) illustrated that the identification of the correct position to install the BF wells is a critical factor for the successful application of the BF technique. However, the major drawback of BF is that it is a site-specific technique, and therefore an extensive investigation must be conducted to determine the site's viability for BF application. Sandhu (2015) proposed a four-stage investigation plan for site selection; this plan can be described as follows: (i) a preliminary evaluation of the potential sites by conducting field studies to collect information on the hydrogeological and hydrological properties of the water systems and collecting samples from groundwater and surface systems: (ii) an in-depth assessment of the potential sites to identify the appropriate locations for installing the BF wells, to determine the groundwater elevations at the investigated areas and to construct monitoring wells; (iii) determination of the hydrological parameters of the aquifer and monitoring of the surface water levels and quality; and (iv) development of an analytical or numerical model to estimate travel time and determine the bank-filtrate proportion in the total water pumped.

The objective of this study was to analyse the performance of BF in Aswan City (Egypt) as an example of an arid climate region. The study was conducted in three phases including (i) the development of a hydrological model for the study area to assess the appropriate locations for BF-well installation and propose different scenarios to manage the BF fields under different environmental conditions; (ii) a water quality assessment - the quality characteristics of the surface water resources and observation wells were monitored for an extended period to predict the bank-filtrate quality; and (iii) an economic analysis conducted to compare the BF costs systems with existing conventional surface water treatment systems. Ultimately, guidelines to facilitate the application of BF in Egypt and countries with a similar hydrological regime were proposed.

6.2 STUDY AREA

Aswan City is located in the south of Egypt on the eastern bank of the NR between 32° 53′ and 32° 56′ E longitudes and 24° 01′ 30″ and 24° 04′ 30″N latitudes (Figure 6.1). The area under investigation was bounded from the east and west by the basement rocks and the NR, respectively. It covered approximately 19.43 km^2, with a maximum width of approximately 4.5 km. The area has a dry climate, without rain, except for one event every 10 to 15 years (Selim et al., 2014). The maximum and minimum air temperatures are 49.5 °C and 27.3 °C in summer, and 30 °C and 15 °C in winter, respectively, and the average annual temperature is 26 °C. The relative humidity in May is the lowest (18%); the maximum occurs in December (40.3%). The ground elevation of the study area ranged from 88–211 m above sea level with an average of 112 m. The study area included three geological units (basement rocks, Nubian Sandstone, and Quaternary sediments). The Quaternary sediments consist of sands, gravels, and clays of the Pleistocene age (Hamdan et al., 2011). In the southern region of the study area, the aquifer is composed mainly of fine to medium sand intersected by thin intercalations of clay; the thickness of these intercalations typically increases from the south to the north. The study area was bounded from the west and east boundaries by complex Precambrian igneous and metamorphic rocks, mainly of granites and schists. Nubian Sandstone strata overlaid the basement rocks with a thickness ranging between 20 and 85 m (Misahah al-Jiyulujiyah et al., 1954; Selim et al., 2014). This study area included 76 pumping wells; however, the majority of these were not in operational mode. Currently, 11 wells pump approximately 23700 m^3/day for irrigation, industrial, and drinking purposes.

6.3 RESEARCH METHODS

6.3.1 Hydrological model

A hydrological model was developed to simulate the current situation of the Aswan's aquifer and propose different scenarios to manage the BF technique.

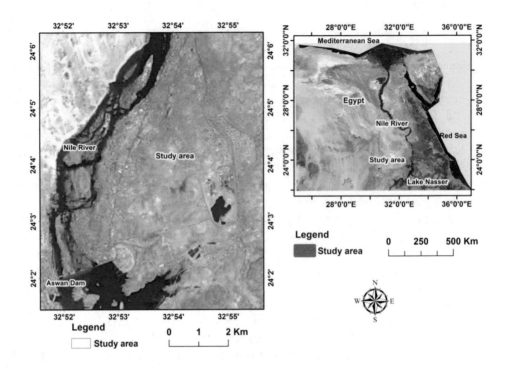

Figure 6.1. Study area location in Egypt (right) and detailed features (left)

Data preparation

This study focused on the development of a 3-dimensional groundwater flow model to characterise the groundwater flow system and levels in the area adjacent to the NR at Aswan City, Egypt, using the coupling of MODFLOW (finite difference code) (McDonald et al., 1988) and Geographic Information System (ArcGIS) (ESRI, 2011). The developed model was used to identify the current hydrological situation of the aquifer, define the proper positions to install BF wells, and assess the effects of design and operation conditions on the efficiency of BF systems. A geodatabase for the Nubian Sandstone Aquifer Aswan was developed from different data sources using ArcGIS. The model data included model geometry, river stage levels, cross sections, pumping test, observation heads (42 observation wells), and borehole data. A digital elevation model (DEM) was derived from the SRTM-3 (Shuttle Radar Topography Mission) (Farr et al., 2000). The geometric surfaces, initials, and the hydraulic parameters were developed using ArcGIS as point data features with their spatial references and appropriate attributes.

This dataset was interpolated using Surfer software and a kriging technique with a proper variogram model, and prepared as input for the model development (Cressie, 1990).

Model development

In MODFLOW, the aquifer was discretised using an array of finite different cells and nodes. The study area was simulated horizontally with a grid mosaic of 71 rows and 90 columns with a cell dimension of 100×100 m resulting in 3,014 active cells. The model boundaries were identified as follows (Figure 6.2). i) The river boundary: the study area was bounded from the west and south by the NR and Aswan Dam Lake (ADL). These natural boundaries were simulated in MODFLOW using the River package. (ii) The aquifer was bounded from the north and east with basement rocks where the lateral

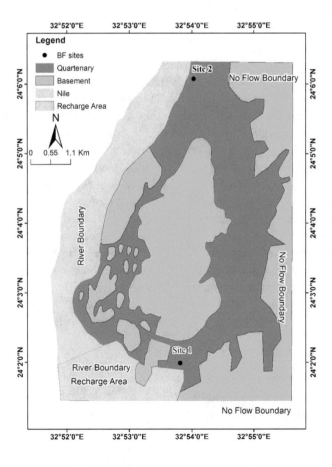

Figure 6.2. . Geology units and model boundary conditions

125

groundwater flow was negligible or non-existent, and thus these outer features were considered as no-flow boundaries. The aquifer base was assigned as a no-flow boundary (basement rocks).

The aquifer was mainly recharged from ADL (the lake confined between the AHD and Aswan dam). Areal recharge from the surface was considered negligible as no or minimal rain occurs yearly in the study area. Based on the available hydraulic heads, the general groundwater flow direction in this area is from south to north corresponding to the NR flow (Selim et al., 2014). The outflow of the aquifer occurs at the northwest of the model area where the water flows from the aquifer into the NR. The NR conductance was estimated for a bed thickness of 1.6 m and hydraulic conductivity of 0.0004 m/s (Ray et al., 2002).

The pumping rates from the production wells were simulated as constant values. Public drinking water network leakage was represented with injection wells with positive charge and capacities based on the data provided by the Aswan Water and Wastewater Company, then validated during the calibration process. The unconfined aquifer was simulated with one layer with variable thickness starting from 130 m in the south and decreasing gradually in a northerly direction. This permeable layer was composed mainly of sand and gravel and demonstrated insignificant horizontal variations in its hydraulic parameters along the scale of the modelled area. Moreover, it was assumed to be internally homogenous anisotropic with equal hydraulic conductivity (K) in the X and Y directions ($K_x = K_y$), and one order less in the Z direction ($K_z = K_x/10$), this is in agreement with other modelling studies that have been conducted in upper Egypt (Bartak et al., 2014; Shamrukh et al., 2005). The $K_{x,y}$ values were assigned as 0.001 m/sec based on the aquifer test and laboratory analysis estimations. The hydraulic parameters, including transmissivity, storage coefficient, and leakage rate of the porous layer were estimated from the pumping tests conducted during this research and were consistent with the values determined by different authors such as Hamdan et al. (2011). These hydrological parameters were calibrated during the simulation period.

Model Calibration

The model was primarily developed and executed under steady-state conditions using the initial estimates of the hydraulic parameters (e.g., hydraulic conductivity). To reproduce the preliminary configuration of the aquifer water table prior to pumping, the hydraulic parameters and stresses were adjusted by a trial and error technique. A calibrated output hydrological map (2008) was used as an initial head map to execute and calibrate the model in the transient condition. Although the steady-state results were reasonable, the steady-state condition was not proven completely. Therefore, a "warming-up period" technique was used in this research to avoid the accumulative error that could appear during the calculation process due to inaccurate initial conditions imposed on the model software (El-Zehairy, 2014). The transient model was initiated with a warming period (365 days). Then, for another five years (from January 1st, 2009 to December 31st, 2013), the model was executed and calibrated under the stress of eight pumping wells and 33 observation wells with variable hydraulic heads, distributed in the study area. The simulation time was discretised into daily stress periods; each had two-time steps. The model was executed repetitively and recalibrated until field-observed values were matched with the modelled values within an acceptable level of accuracy. Then, the model was subsequently executed and validated for a further five years (2014–2017) as a validation period.

The calibrated results demonstrated an acceptable agreement between the modelled and observed groundwater heads with $R^2 = 0.90$. The mean absolute error (MAE) was 0.37 m, and the root mean square error (RMSE) was 0.47 m (Figure 6.3a), which indicates reasonable model performance. The validated results also indicated that the modelled and observed heads were well aligned ($R^2 = 0.90$, MAE = 1.02 m, and RMSE = 1.31 m) (Figure 6.3b). Moreover, observed and simulated heads in both cases were virtually aligned around the mean of the observed heads.

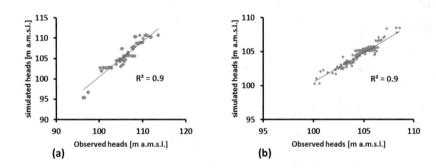

Figure 6.3. Model simulated vs field observed water level: (a) calibration period and; (b) validation period

6.3.2 Development of BF management scenarios

A calibrated hydraulic simulation was used to optimise the hydraulic capacity of the BF system and assess the influence of the design conditions including well spacing, distance from the well to the river, and pumping capacity, and to determine the effect of surface water level on the efficiency of the BF technique. In this research, two proposed BF sites

127

(Site 1 and 2, Figure 6.2) were examined. Site 1 was located at the recharge region of the aquifer (specifically at longitude 24° 01′ 56″ N and latitude 32° 53′ 48″ E); this site was 100 m farther from the lake reservoir confined between the Aswan Dam and AHD. Site 2 was located at 24° 06′ 05″N, 32° 54′ 01″E. This site was characterised by its proximity to an urban area.

The best options were determined based on three criteria: (i) appropriate travel time, (ii) greater bank-filtrate share, and (iii) less drawdown. The MODPATH particle-tracking code (McDonald et al., 1988; Pollock, 1989) was used to estimate the travel time of water particles to the BF pumping wells. MODPATH is a 3-dimensional tool designed for collaborating with MODFLOW to determine the advective transport of particles that mimic pollutants (or tracers). During this research, a set of particles weres identified at the interface between the surface water system and adjacent aquifer. Then, the MODPATH code was applied in the forward function to simulate the migration of the water particles toward the wells. Abdel-Fattah et al. (2008) illustrated that river water particles that flow into the pumping wells following straight trajectories (meridian pathlines) require less time to reach the wells than particles on angled paths (angled pathlines). In this research, the travel time along the meridian and angled lines were estimated and used as an approximation for the minimum and maximum travel times. Grützmacher et al. (2010) stated that the elimination of cyanobacteria toxins during the BF process requires at least ten days of residence. Wintgens et al. (2016) demonstrated that a subsurface travelling time of 50 days is adequate to remove pathogens and provide high-quality drinking water. Maeng et al. (2010), conversely, found a negative relationship between the travel time and redox potential of the bank filtrate. This suggests that longer travel increases the potential for environmental anaerobic conditions. This enhances the reduction of undesirable and toxic elements (e.g., Fe, Mn, and As), and consequently has an adverse effect on the bank-filtrate quality. Based on these assumptions, a travel time of 10 to 50 days was regarded as acceptable.

The ZoneBudget (ZONBUD) code (Harbaugh, 1990) was applied to estimate the bank-filtrate share for each hydraulic simulation. ZONBUD is post-processing code that facilitates the determination of a sub-region water budget based on the MODFLOW flow model results. First, a set of fictitious particles were placed at the locations of the extraction wells. Then, the MODPATH code was execute in the backward mode to delineate the pathlines of these particles from the aquifer and surface water system toward the abstraction wells and determine the size of each zone. Then, the ZONBUD code was applied to estimate the contribution of the ambient groundwater and infiltrated water to the total pumped water.

6.3.3 Water quality characterisation

To predict the quality characteristics of the bank filtrate, water samples were collected from a pumping well (BF1) placed at 24 02′ 28″N and 32° 54′ 35″E near the first potential BF site and 1.2 Km from the surface water system. A further set of samples were collected from an observation well (BF2) situated at 24° 06′ 03″ N and 32° 54′ 08″E near the second potential BF field (Site 2) and 600 m from the Nile. Moreover, samples were collected from the surface water systems (NR and ADL) and a groundwater well (GW) (located at 24° 03′ 26″ N and 32° 54′ 33″E) to assess the efficiency of the BF process. The samples were collected regularly for one year (from January 2017 to December 2017), filtered using a 0.45-μm membrane filter (Whatman, Dassel, Germany) and analysed. The physical parameters (temperature, electric conductivity, pH, and turbidity) were determined at the field using portable (HACH, USA) instruments. The main inorganic parameters were quantified using an ion chromatograph (881 Compact IC pro, Metrohm, Swiss); the metals were analysed by ICP-OES (Optima 8300 from Perkin Elmer Company, USA).

The bulk organic concentration of the raw and infiltrated water was determined using a total organic carbon (TOC) analyser (TOC-VCPN (TN), Shimadzu, Japan). The water absorbance in the ultraviolet range at the 254 nm wavelength (UV_{254}) was used as a predictor for the aromaticity and potential formation of trihalomethane during the treatment process (Maeng et al., 2019). The UV_{254} [cm^{-1}] absorbance was monitored using a UV/Vis spectrophotometer (UV-2501PC Shimadzu). Then, the specific ultraviolet absorbance ($SUVA_{254}$) [L mg^{-1} m^{-1}] was calculated by dividing the UV254 absorbance [m^{-1}] by the dissolved organic carbon (DOC) concentration [mg/L]. The measurements were conducted at the laboratories of the Egyptian Holding Company for water and wastewater.

Fluorescence excitation-emission measurements (F-EEM) were conducted following the procedures described in (Abdelrady et al., 2019) to classify the bulk organic matter into different constituents. The fluorescence measurements were performed once at the IHE- (Delft, Netherlands) using a Fluoromax-3 spectrofluorometer (HORIBA Jobin Yvon, Edison, NJ, USA). The fluorescence characteristics were determined at an excitation wavelength (λ_{ex}) between 240–452 nm (interval = 4 nm); the emission wavelengths (λ_{em}) were from 290–500 nm with an interval of 2 nm. In this study, three fluorescent peaks were identified at definite excitation and emission wavelengths representing three different organic substances. These peaks included primary humic-like P1 (λ_{ex}= 250–260 nm and λ_{em} = 380–480 nm), secondary humic-like (λ_{ex}= 300–370 nm and λ_{em} = 400–500 nm), and protein-like (λ_{ex}= 270–280 nm and λ_{em} = 320–350 nm) (Coble, 1996; Leenheer et al., 2003). To gain insight into the organic characteristics of the bank filtrate and raw water, the fluorescence indices (Humification index (HIX), fluorescence index (FIX), and

129

biological index (BIX)) were estimated following the equations presented in (Gabor et al., 2014).

The percentage of the infiltrated water from the surface water systems captured by the two BF wells was determined using conservative chemical parameters (e.g. electrical conductivity and chloride) based on the following equation (Lamontagne et al., 2015; Zhu et al., 2020):

$$BF\% = \frac{C_{BF} - C_{GW}}{C_{SW} - C_{GW}} \times 100 \tag{1}$$

where C_{BF}, C_{SW}, and C_{GW} are the concentrations of the conservative parameter in the BF well, surface water, and native groundwater, respectively.

6.3.4 Cost analysis

The extended application of BF in developing countries is highly dependent on its capacity to provide a sufficient quantity and quality of drinking water. Nonetheless, the economic cost is also a decisive factor that must be considered. Two methods (Net Present Value (NPV) and Payback Period (PBP)) were used in this research to evaluate the economic feasibility of using the BF technique in Aswan City compared to other existing treatment techniques. PBP calculates the minimum time (in years) required to recover the total investment cost. This can be calculated as follows:

$$PBP = \frac{Total\ capital\ cost}{annual\ income - annual\ outcome} \tag{2}$$

The NPV is used to assess each project's profitability by offsetting all future income and expenses to the present (Equation 3):

$$NPV = \sum_{n=1}^{N} \frac{B - C}{(1+r)^n} - I_0 \tag{3}$$

Where B and C are income and expenses of each year, r is the discount rate, and I_0 is the capital cost. Whereas, n is the project's lifetime (in years), and has been assigned as 25 years (Bonton et al., 2012).

6.4 RESULTS AND DISCUSSION

6.4.1 Aswan aquifer model (Current situation)

The model was developed to simulate the current situation (2009–2017) of the Aswan city aquifer and assess the influence of the proposed stresses on the interaction between the surface water and groundwater systems and consequently on the water quantity of the BF wells. The ultimate goal was to provide different scenarios to manage the bank fields under different hydrological conditions. The results revealed that the ground and infiltrate water flow from the south to the north direction, paralleling the NR. The total water penetration from the ADL into the aquifer was highly dependent on the lake and river stages and aquifer's water head; its value ranged from 683710 m^3/day in summer to 551030 m^3/day in winter. The groundwater had the same general flow path as the NR from the south to the north. Conversely, discharging water from the aquifer toward the NR (baseflow) was the main outflow component and occurred in the northern region of the aquifer. The total water discharge into the surface water system ranged from 528410 to 603500 m^3/day. The aquifer exhibited a high capacity to store the water. In 2009, 40 productive wells were shut off, and consequently, the groundwater head increased by 1–3 m during the period from January 2009 to July 2013 (Figure 6.4a and b). The groundwater level increase resulted in the creation of ponds in the low-lying areas, with depths ranging from 8–15 m. These ponds have detrimental environmental impact as they endanger the public health and city's infrastructure (Selim et al., 2014). Therefore, the government was obligated to reoperate wells and pump water from the aquifer. In 2013, eight abstraction wells were used to pump 17280 m^3/day from the aquifer. Consequently, the groundwater level decreased by 0.3 m and 0.5 m by the end of 2013 and 2017, respectively (Figure 6.4c, d, and e). Hence, there is a requirement to install BF wells to produce high quality drinking water and promote the decreasing of the groundwater level.

6.4.2 Bank filtration management scenarios

Different hydraulic simulations were performed to assess the influence of the well design parameters and identify the optimum operating conditions for managing the BF technique at two different sites in Aswan City.

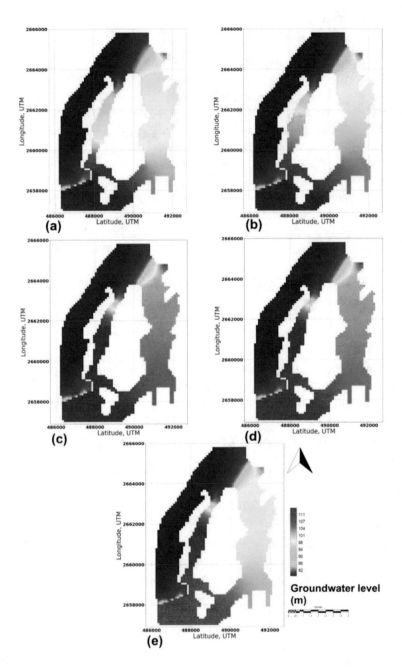

Figure 6.4. Simulated groundwater levels of Aswan aquifer for years (a) 2008, (b) 2009, (c) 2013, (d) 2014, and (e) 2017.

Effect of number of wells and pumping rate on BF performance

Three scenarios were proposed to evaluate the influence of the number of wells and pumping rate on the BF performance:

- Scenario 1 (effect of the number of wells): Four simulations were conducted based on the number of wells; 5, 10, 15, and 20 wells, where the production capacity was the same (35000 m^3/day) in each simulation. The production capacity was divided equally on the number of wells, so that, in each simulation, each well has the same puming rate.
- Scenario 2 (effect of pumping rate): Three simulations were conducted based on the pumping rates (35000, 17770, and 70000 m^3/day). The number of wells in each simulation was constant (10 wells).

- Scenario 3 (effect of increasing the number of wells and pumping rate simultaneously): A different groups of 5, 10, 15, and 20 wells were simulated with production capacity of 17500, 35000, 52500 and 70000 m^3/day, repectively.

For all the scenarios the wells were aligned parallel to surface water system at 100 m far from the surface water syatem, and the space between the wells was 50 m.

The Scenario 1 results (Figure 6.5a) illustrated that increasing the number of wells, while maintaining the production capacity constant (35000 m^3/day), could influence the BF efficiency. After an operation period of 365 days, the bank-filtrate share was increased from 73 to 97% and from 65 to 70% when the number of extraction wells increased from 5 to 20 at Sites 1 and 2, respectively. Nonetheless, no noticeable influence on the drawdown was observed. The drawdown was increased marginally from 1.4 to 1.5 m at Site 1 and decreased from 5.6 m to 3.7 m at Site 2. Conversely, travel time along the meridian pathline increased from 13 to 19 days and 12 to 23 days when the number of wells were increased from 5 to 20 in the two planned areas. However, travel time along the angled pathlines declined from 60 to 48 days and 72 to 40 days. Hence, it can be deduced that installing a low number of wells (less than 10) at the proposed sites could increase the risk of developing an anaerobic environment during the infiltration process.

The results obtained from Scenario 2 (Figure 6.5b) demonstrate that increasing the pumping rate considerably influenced the BF efficiency. For Site 1, a double increase in the abstraction rate of the extraction wells decreased the bank-filtrate share of the pumped water by 5–10% and dropped the drawdown by 1–1.5 m. Furthermore, the travel times were 95, 30, and 30 days for the particles travelling along the angled pathlines and 21, 11, and 7 days for the particles tracking the meridian pathlines when the abstraction capacity was 17500, 35000, and 70000 m^3/day, respectively. The same findings were observed at Site 2, that is, that increasing the production capacity of the wells could increase the drawdown and bank-filtrate share values and reduce the travel time. As the pumping

133

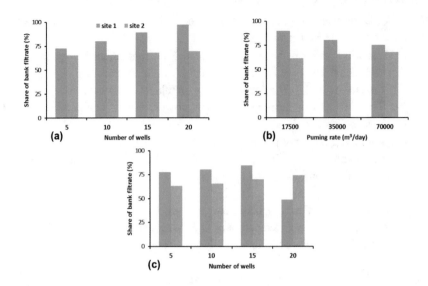

Figure 6.5. Effect of number of wells and pumping rate on bank-filtrate share (BF %): (a) Scenario 1, (b) Scenario 2, and (c) Scenario 3.

capacity increased by 17500 m^3/day, the bank-filtrate share improved on average by 2%. Moreover, the drawdown estimated was 2.1, 5.2, and 11.6 m at an abstraction capacity of 17500, 35000, and 70000 m^3/day. Similarly, the travel time was decreased from 40, to 15, and to 8 days as the production capacity increased from 17500, to 35000, and 70000 m^3/day, respectively. This inferred that a higher production capacity (> 35000 m^3/day) could cause a severe increase in drawdown values and reduce travel time to a level (< 10 days) that influences the efficiency of the BF processes at the proposed BF sites.

The outcomes of Scenario 3 (Figure 6.5c) revealed that increasing the abstraction capacity and number of wells simultaneously at Site 1 had a notable influence on the bank-filtrate share of the pumped water. It estimated that the river-infiltrated water contributed to the pumped water with 78, 80, and 84% when the number of abstraction wells was 5, 10, and 15, respectively. However, this value decreased to 49% when the number of wells increased to 20, with a total production capacity of 70000 m^3/day. This implies that an over-pumping process induces the ambient groundwater to flow toward the BF wells and thus promotes the groundwater's contribution percentage to the produced water and reduces the share of the bank-filtrate percentage. This is in agreement with the drawdown results, which indicated a gradual decreased behaviour with an increased the number of wells. The groundwater table was depleted by approximately 0.7 m when the number of wells increased from 5 to 20. Conversely, it was observed that doubling the number of wells and their pumping intensities concurrently enhanced the bank-filtrate share by approximately 4% at Site 2 and reduced the groundwater level by 1.5 m on average. This procedure has also been demonstrated to have a substantial effect on the travel time of the infiltrating water. The results displayed indicate that it required 10 and 20 days for the water particles following the meridian pathlines to reach the pumped wells in Sites 1 and 2 when the number of wells was 15 and 20, respectively. Whereas the particles following the angled pathlines converged at the pumping wells after 100 days of infiltration time when five wells were simulated. This infers that both low and high pumping procedures adversely influence the travel time and ultimately reduced the BF efficiency of at the two proposed areas.

From all scenarios, from a hydrological perspective, it was determined that the BF performance at Site 1 was superior to that at Site 2. For a production capacity of 35000 m³/day (with a number of wells ranged between 5-20 and at the spacing of 50 m), the contribution of river infiltrated water to the total pumped water ranged between 73 and 97% at Site 1, whereas it ranged between 65 and 70% at Site 2 under the same design conditions. Further, the maximum estimated drawdown was 1.8 and 5.6 at Site 1 and 2, respectively. It was also noticed that the production capacity of the wells had a greater influence than the number of wells. Increasing the number of wells, maintaining the total production capacity constant, could improve the bank-filtrate's share without a major change in travel time and drawdown. The bank-filtrate proportion exceeded 80% at Site 1 and 65% at Site 2 when ten or more extraction wells with an abstraction capacity of 35000 m³/day were installed. Furthermore, the travel time varied between 10 and 50 days, which is sufficient to enhance BF efficiency. Higher production capacity, however, could shorten the travel time to the limit, reducing the treatment efficiency of the BF processes.

Effect of distance of the Well from river on BF performance

Several hydraulic simulations were developed to analyse the influence of the distance between the surface water system and abstraction wells on the BF performance at the two potential sites. The production wells were modelled at different distances (i.e., 50, 100, and 200 from the river) parallel to the surface water systems. The effect was assessed at different well-pumping rates (17500, 35000, and 70000 m³/day). The results (Figure 6.6) indicated that the distance of the wells to the river did not have a remarkable influence on the bank-filtrate share and drawdown. After 365 days of operation, the bank-filtrate percentage in the abstraction wells (with a production capacity of 35000 m³/day) situated at a distance of 50 m from the river was 81% and 66%, and the drawdown was 1.7 m and 4.4 m at Sites 1 and 2, respectively. In comparison, placing the BF wells at 200 m from the river changed the proportion of infiltrated water in the total pumped water by 1% and depleted the groundwater table by 0.3 and 2.2 m at the two proposed sites, respectively. The same finding of a minor effect of the distance between the BF well and river on the

135

bank-filtrate share and drawdown was determined at other pumping rates. Conversely, travel time was the most influenced parameter of changing the distance from the wells to the river. When the wells were positioned at 200 m from the river, and the pumping rate was low (1750 m³/day), the water particles converged at the abstraction wells at a time ranging between 135–195 days at Site 1 and 83–110 days at Site 2. As the pumping rate doubled, the travel time at both sites decreased to 72–150 and 43–65 days, respectively. Consequently, there is a high potential of the infiltration environment becoming anaerobic under these hydrological conditions. Conversely, the proximity of the wells to the surface water system reduced the time of interaction between the soils and infiltrated water and therefore influenced the efficiency of the treatment processes. At a distance between the wells and river of 50 m, the travel time varied between 13–35, 7–35, and 3–17 days at pumping rates of 1750, 3500, and 7000 m³/day, respectively. This indicates that low pumping rates should be employed when the BF wells are close (50 m or less) to the surface water system at the proposed sites.

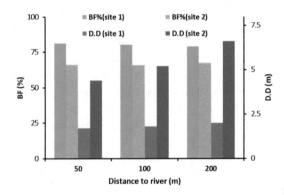

Figure 6.6. Impact of the distance between BF wells and surface water system on the bank filtrate share (BF %) and drawdown (D.D) at the two potential BF sites of Aswan City

Effect of well spacing on BF performance

The space between the wells has a significant influence on the level of the groundwater and its interaction with the surface water system and subsequently affects the efficiency of the BF system. In this research, two well spacing options (i.e., 25 and 50 m) were assessed under different design and operation conditions, including different pumping rates and distances to the river. From a hydrological perspective, it was observed that the space between the pumping wells had a positive effect on the BF performance at the two investigated sites (Table 6.1 and Table 6.2).

Table 6.1. Effect of well spacing on BF performance parameters (i.e., bank-filtrate share BF%, drawdown D.D, and travel time) for wells placed at different distances to surface water system.

Well spacing (m)	Distance to river (m)	Site 1			Site 2		
		D.D (m)	BF%	Travel time (day)	D.D (m)	BF%	Travel time (day)
	50	1.3	91	12–25	3.4	70	15–30
50	100	1.6	92	22–47	4	71	25–35
	200	1.8	92	103–175	5.5	73	53–70
	50	2	81	7–25	6.6	66	7–30
25	100	1.8	80	11–50	5.2	66	18–40
	200	2	79	72–150	6.6	67	43–65

Table 6.2. Effect of well spacing on BF performance parameters (i.e., bank-filtrate share BF%, drawdown D.D, and travel time) for wells operating at different pumping rates (m³/day).

well spacing (m)	Pumping rate (m³/day)	Site 1			Site 2		
		D.D (m)	BF%	Travel time (day)	D.D (m)	BF%	Travel time (day)
	1750	1.2	99	36–90	2.1	63	47–60
50	3500	1.6	92	22–47	4	71	25–35
	7000	1.8	84	14–45	5.2	75	12–17
	1750	1.3	90	21–95	2.6	62	40–80
25	3500	1.8	80	11–50	5.2	66	18–40
	7000	2	75	7–40	8.7	68	25–35

In all hydraulic simulations, it was observed that an increase in space between the wells from 25 to 50 m could enhance the proportion of infiltrated water in the total produced water by approximately 10% at both sites. Furthermore, it reduced the drawdown, thereby prolonging the travel time. This effect can be attributed primarily to the expansion of the radius of influence of the wells when they are close (Sharma et al., 2012). When the well spacing increased from 25 to 50 m, the drawdown reduced by an average of 0.2 m at Site 1 and 3 m at Site 2. In comparison, the travel time at both sites increased by, respectively, 5–25 and 5–10 days. However, inappropriate travel time was recorded when the wells were installed 200 m from the surface water system with a low pumping rate (1750 m^3/day), as reported in earlier sections.

Effect of river stage on BF performance

Several hydraulic simulations were performed to determine the influence of decreasing the river stage on BF performance at Aswan City. Within these simulations, surface water levels were reduced by 0.5, 1 m and 1.5 m as the worst scenario. The results (Figure 6.7) indicated that reducing the surface water level could decrease the bank-filtrate contribution to the total pumped water, particularly at the onset of the BF wells' operation. However, after a period of approximately 100 days, the effect of surface water levels became minor. When the river stage decreased (ΔR.S.) by 1.5 m, the bank-filtrate share at Site 1 and Site 2, respectively, was reduced by 14% and 5% after a 10-day operating period. However, after 90 days, the variance in the bank-filtrate share due to the decline of the river stage was less than 2%. Lowering the surface water level reduces the water table at the BF field, diminishes the hydraulic gradient between the wells and surface water stages, and ultimately reduces the flow velocity of the infiltrated water to the wells. Decreasing the river stage in the model by 1.5 m triggered a 1.2 m reduction of the drawdown at Site 1 and 1.7 m at Site 2. This was followed by an 11–12-day and 2–4-day decrease in the travel time at both sites, respectively. A decrease of the river stage by 0.5 m reduced the subsurface water table by 0.5 m at Site 1 and 0.4 m at Site 2 and lengthened the travel time by 4 and 7 days at the two sites, respectively. Therefore, it can be concluded that the continued functioning of the BF wells can minimise the influence of the reduction of the surface water level and thus improve the BF efficiency.

Bartak et al. (2014) investigated the BF performance at the Dishna site along the NR and reported that the intermittent operation of the BF wells was one of the major drawbacks reducing the efficiency of the BF technique. This study demonstrated that the bank-filtrate share in specific BF wells did not exceed 10% after construction and operation of BF abstraction wells for 1.5 years, primarily due to the 8-hour daily interruption in the operation of the wells. Moreover, the water pumped did not meet drinking water requirements. Therefore, it can be concluded that the continuous operation of BF wells is a prerequisite for the successful application of the BF technique in Egypt.

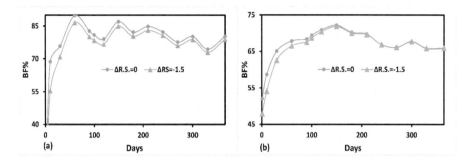

Figure 6.7. Influence of declining river stage on bank-filtrate share (BF%) at two proposed sites: (a) Site 1 and (b) Site 2 at Aswān City (Egypt).

6.4.3 Bank-filtrate chemistry

The chloride was used as conservative elements to estimate the percentage of infiltrated water from the surface water systems to the total pumped water at the two BF wells. The average chloride concentrations for NR, ADL, BF1, BF2, and GW were 6.7 ± 1.7, 7 ± 2, 16.5 ± 3, 38 ± 9, and 56 ± 7 mg/L, respectively. Therefore, the bank-filtrate share for BF1 and BF2 was estimated to be 81% and 36%, respectively.

The physical and chemical characteristics of the raw, infiltrated, and groundwater at the study area are summarised in (Table 6.3 and Table S6.1–S6.3). The minimum and maximum temperatures were 17.8 and 28.6 °C for the surface water systems, respectively, and 16.8 and 25.7 °C for the BF wells, primarily due to the low buffer heat capacity of the soil and short distance between the surface water sources and wells. Moreover, it was observed that the banks acted as a robust barrier to the elimination of suspended matter. The turbidity was low (< 0.3 NTU) at the BF wells, regardless of the corresponding value at surface water systems.

The concentrations of major cations (Na, K, Ca, and Mg) and anions (Cl, SO_4, and HCO_3) demonstrated increased behaviour during the infiltration process. For example, the average concentrations of Ca were 24 ± 3 and 19 ± 5 mg/L for the NR and ADL surface water systems and increased to 36 ± 5 and 55 ± 8 mg/L for the BF1 and BF2 wells, respectively. Similarly, an increase in the concentrations of heavy metals was detected at

139

the BF wells. Fe, Mn, Al, and Cu were not detected in the surface water systems, yet were 20 ± 3, 14 ± 5, 837 ± 116, and 14 ± 7 µg/L at BF1 and 145 ± 2, 211 ± 27, 511 ± 32, and 24 ± 9 µg/L at BF2, respectively. This is mainly attributed to the (i) dissolution of minerals during the filtration process and (ii) mixing of infiltrated water with the contaminated native groundwater (Abdelrady et al., 2020). The dissolved oxygen ranged between 5.1–7.3 mg/L at the surface water sources and decreased to 2.7–5.4 mg/L at the BF wells. It is therefore expected that microbial reduction does not have a major role in increasing the concentration of metals in the pumped water at the BF wells. However, the concentrations of these elements in the pumped bank filtrates did not exceed the threshold levels of drinking water quality guidelines proposed by WHO (2011) and therefore they do not pose a risk to human health. Similarly, other toxic metals (As, Cd, Co, Ni, Pb, and Zn) were not found in either the surface water systems or bank filtrates.

BF is recognised as an effective technique to reduce nutrient concentrations significantly (Maeng et al., 2019). However, higher concentrations of nitrogen and phosphorus were found in the BF wells relative to the concentrations of the surface water bodies during this study. The concentration of ammonia was less than the detection limit at the surface water sources; however, its concentration increased to 0.54 mg/L at BF1 and 0.73 mg/L at BF2. Similarly, the average concentration of phosphate was 0.05 ± 0.02 and 0.03 ± 0.01 mg/L at the NR and ADL systems and increased to 0.1 ± 0.04 and 0.14 ± 0.05 mg/L at the BF1 and BF2 wells, respectively. The principal reason for this increase is the mixing of infiltrated water with contaminated groundwater (Hamdan et al., 2011; Selim et al., 2014).

Organic matter is the driving force behind the process that occurs during the BF process. The bulk and constituents of the organic matter were monitored during this study using different analytical techniques. BF is regarded as an effective technique for reducing the organic matter concentration (Dragon et al., 2018; Nagy-Kovács et al., 2019). Maeng et al. (2008) conducted column studies to simulate BF processes and demonstrated that more than 50% of the organic matter was removed during the top 50 cm of the column. However, in this research, a marginal increase in the concentration of bank-filtrate organic matter was observed. The concentration of organic matter in the NR and ADL surface water systems was 3.9 and 3.6 mg/L, whereas the concentrations in the pumped water of the BF1, BF2, and GW wells were 4.3, 4.9, and 5.5 mg/L, respectively. Moreover, SUVA values of NR, ADL, BF1, BF2, and GW were 1.56, 1.33, 2.16, 2.04, and 2.73 L/mg.m, respectively, which indicates that the bank filtrate and ambient groundwater had relatively higher aromatic characteristics than the surface water sources. This observation could be attributed to the (i) dissolution or desorption of soil organic matter into the filtrate water, which is significantly increased at high temperatures (Abdelrady et al., 2018; Schoenheinz, 2004), (ii) accumulation and subsequent degradation of particulate organic matter (e.g., Phytoplankton) during the filtration process that augments the organic concentration in the bank filtrate (Grünheid et al., 2005), and (iii) effect of mixing the infiltrated water with the ambient groundwater, which has a higher concentration of organic matter. The phenomenon of enrichment of the organic content of the bank filtrate was also reported at BF wells along the Lot River (France) and Lagoa do Peri Lake (Brazil) (Kedziorek et al., 2008; Romero-Esquivel et al., 2017).

Table 6.3. Quality parameters of surface waters, bank filtrates, and groundwater sources.

Parameter	Unit	Egyptian Standards	River Nile	Aswan Dam Lake	BF (Site1)	BF (Site2)	GW
pH		6.5–8.5	8.2 ± 0.6	7.9 ± 0.5	8.41 ± 0.3	8.3 ± 0.5	8.4 ± 0.3
Conductivity	μs/cm	-	234 ± 12	225 ± 8	288 ± 25	376 ± 38	438 ± 26
turbidity	NTU	1	1.5 ± 0.6	1.2 ± 0.7	0.23 ± 0.1	0.28 ± 0.1	0.75 ± 0.2
NH_4^+	mg/L	0.5	n.d.	n.d.	0.42 ± 0.2	0.61 ± 0.1	0.64 ± 0.2
NO_2^-	mg/L	0.2	0.14 ± 0.1	0.1 ± 0.08	0.05 ± 0.02	0.28 ± 0.1	0.37 ± 0.1
NO_3^-	mg/L	45	2.63 ± 0.4	2.22 ± 0.6	0.71 ± 0.1	5.8 ± 1.3	4.3 ± 0.8
Fe	μg/L	300	n.d.	n.d.	20 ± 3	145 ± 21	330 ± 37
Mn	μg/L	400	42 ± 9	n.d.	14 ± 5	211 ± 27	432 ± 41
DOC	mg/L	-	3.90	3.60	4.30	4.90	5.50
SUVA	L/mg.m	-	1.56	1.33	2.16	2.04	2.73
P1	R.U.	-	0.32	0.20	0.41	0.74	1.10
P2	R.U.	-	0.12	0.09	0.14	0.25	0.46
P3	R.U.	-	0.23	0.22	0.11	0.14	0.19

MDL: minimum detection limit; n.d.: not detected; R.U.: Raman Unit

The organic fluorescence characteristics of the surface water, bank filtrates, and groundwater were determined to provide insight into the organic composition of the water sources.. The protein-like components were considerably attenuated during the filtration process. The fluorescence intensity results (Table 6.3) indicated that the protein-like compounds were reduced by 33–55% in the bank filtrates compared to their intensities in the surface water systems. Therefore, it can be concluded that BF enhances the biological stability of pumped water (Maeng et al., 2019). Conversely, bank filtrates had a relatively higher content of humic compounds than surface water sources. This can be mainly attributed to the mixing effect of the infiltrated water with groundwater with high humic content, and the dissolution of soil humic compounds into the infiltrate water as mentioned above. The fluorescence indices verified these findings. The results indicated that the bank filtrates had higher humic content (higher HIX) and lower microbial-derived compounds (lower BIX). The FIX was 1.67 and 1.36 for the NR and ADL surface water systems and decreased to 1.09 and 1.07 for bank filtrates BF1 and BF2, respectively. This infers that the bank filtrates retained a higher content of humic compounds of terrestrial origin than the surface water sources. The increase of humic compounds in the bank filtrates could enhance the formation of trihalomethane compounds during the post-treatment process (i.e., post-chlorination) (Awad et al., 2016). Consequently, the BF design process in Aswan city must focus more on reducing the proportion of groundwater in the overall pumped water. Moreover, a post-treatment process (e.g., coagulation and filtration) might be needed to minimise the humic concentration in the bank filtrate.

6.4.4 Economic analysis

The primary sources of drinking water in Aswan City are conventional water treatment plants (WTPs) and groundwater wells. High-capacity WTP (production capacity ranges between 17,000 m^3/day and 1 million m^3/day or more) employs coagulation, flocculation, sedimentation, sand filtration, and disinfection combination to provide potable water to the public. Low-capacity WTP (compact units, a production capacity ~2160 m^3/day) is mainly used to provide drinking water to small villages. The treatment processes are similar to that of high-capacity WTPs, but on smaller-scale (Wahaab et al., 2019).

BF and GW plants consist of extraction wells and a chlorine-disinfection unit, no post-treatment procedures are presently implemented. The depth of the GW wells ranged from 90 to 110 m. The BF wells, however, are drilled at a depth of about 20-60 m. The existing treatment techniques produce drinking water quality meeting the Egyptian standards.

This section presents an economic comparison between the existing treatment techniques and BF in Aswan city. The capital and operation costs (including chemicals costs) of each treatment technique were estimated based on expenditure and revenue field data for five years (2015–2019) from the Aswan Water and Wastewater Company, Egypt. The water tariff is 2.09 EGP per cubic meter. The operation costs of BF and GW wells were assumed to be the same.

The results (Table 6.4) revealed that NPV and PBP were 8.8 Million EGP and 0.5 years for BF, 8.4 Million EGP and 0.7 years for GW, 0.2 Million EGP and 8.2 years for a low-capacity WTP, and 7.6 Million EGP and 6.3 years for a high-capacity WTP, respectively. This indicates that the BF technique had the greatest NPV and least PBP, and therefore, the BF technique is economically feasible to implement in Aswan City.

Table 6.4. Economic comparison between current treatment techniques and BF in Aswan City, Egypt

	Unit	BF	GW	Low –capacity WTP	High- capacity WTP
Productio n capacity	m³/day	2160	2160	2160	8640
Capital cost	Million EGP/Unit	0.5	0.8	5.0	35.0
Operation and energy cost/ year	Million EGP/Unit	0.5	0.5	1.0	1.8
NPV	Million EGP	6.2	5.9	0.9	7.6
PBP	years	0.5	0.7	8.2	6.3

6.5 CONCLUSIONS

This study examined the application of the BF technique in Aswan City (Egypt). Several hydraulic simulations were developed to assess the influence of environmental conditions and BF design parameters on the BF performance at two potential sites. The bank-filtrate share, drawdown, and travel time, which are the most critical parameters, were examined. Based on the simulation results, the following guidelines were proposed:

- The number of wells and pumping rate influence the BF performance parameters and thus the bank-filtrate quality. The bank-filtrate share exceeded 65% at both sites when 10 wells with a production capacity of 35000 m³/day located 100 m from the river were simulated. The travel time ranged from 10 to 50 days, which is sufficient for the effective removal of majority of the pollutants. However, a decrease in the number of installed wells could reduce the bank-filtrate proportion and increase the travel time that increases the potential for the development of anaerobic conditions in the infiltration zone. Conversely, over-pumping practices could reduce travel time and, as a consequence, the effectiveness of treatment.

- Installing the wells close to the river (≤ 50 m) could shorten the travel time to the limit that impedes the treatment process. To overcome these difficulties, a low production capacity (≥ 17500 m^3/day) must be applied.
- The reduction in the surface water level (by 0.5-1.5 m) has a considerable effect on the BF performance in the onset operation of the wells. However, this effect reduces with the continuous operation of the wells.

Furthermore, water samples were collected from the surface water sources and observation wells placed at the adjacent aquifer to characterise the chemical quality parameters of the water sources and the bank filtrate. The results infer that the bank-filtrate water achieved the standard for drinking water. However, an increase in the organic matter (particularly humic compounds) was observed that could be attributed to the effect of mixing infiltrated water with polluted groundwater or the dissolution of soil organic matter.

Finally, an economic comparison was performed to compare BF and the existing treatment technique in Aswan City. The results revealed that BF has lower payback and NPV than the other techniques, which infers that it has high economic viability.

This study concluded that BF has a high degree of feasibility for application in Aswan City (Egypt). However, certain considerations must be addressed during the design process.

6.6 SUPPLEMENTARY DOCUMENTS

Table S6.1. Physical and inorganic characteristics of the surface waters, bank filtrates and groundwater sources

Parameter	Unit	Egypt standards	River Nile	Aswan Dam Lake	BF (site1)	BF (site2)	GW
pH		6.5-8.5	8.2±0.6	7.9±0.5	8.41±0.3	8.3±0.5	8.4±0.3
Conductivity	µs/cm	-	234±12	225±8	288±25	376±38	438±26
turbidity	NTU	1	1.5±0.6	1.2±0.7	0.23±0.1	0.28±0.1	0.75±0.2
Ca^{+2}	mg/L	-	24±3	19±5	36±5	55±8	70±6
Mg^{+2}	mg/L	-	5±1	7±1	9±2	22±5	19±2
Na^+	mg/L	200	11±3	10±1	11.4±2	26±3	32±5
K^+	mg/L	-	5.3±0.8	4.8±2	5.5±1	7.4±2	8.4±4
NH_4^+	mg/L	0.5	n.d.	n.d.	0.42±0.2	0.61±0.1	0.64±0.2
NO_2^-	mg/L	0.2	0.14±0.1	0.1±0.08	0.05±0.02	0.28±0.1	0.37±0.1
NO_3^-	mg/L	45	2.63±0.4	2.22±0.6	0.71±0.1	5.8±1.3	4.3±0.8
SO_4^{-2}	mg/L	250	12±3	14±4	19±4	33±9	25±8
PO_4^{-3}	mg/L	-	0.05±0.02	0.03±0.01	0.1±0.04	0.14±0.05	0.25±0.08
Cl^-	mg/L	250	6.9±1.7	7±2	16.5±3	38±9	56±7

Table S6.2. The concentrations of heavy metals in the surface waters, bank filtrates and groundwater sources

Parameter	Unit	Egypt standards	River Nile	Aswan Dam Lake	BF (site1)	BF (site2)	Ambient GW
Fe	μg/L	300	n.d.	n.d.	20±3	145±21	330±37
Mn	μg/L	400	42±9	n.d.	14±5	211±27	432±41
As	μg/L	10	n.d.	n.d.	n.d.	n.d.	n.d.
Cd	μg/L	3	n.d.	n.d.	n.d.	n.d.	n.d.
Co	μg/L	-	n.d.	n.d.	n.d.	n.d.	n.d.
Cr	μg/L	50	n.d.	n.d.	n.d.	n.d.	n.d.
Cu	μg/L	2000	n.d.	n.d.	14±7	24±9	32±7
Ni	μg/L	20	n.d.	n.d.	n.d.	n.d.	n.d.
Pb	μg/L	10	n.d.	n.d.	n.d.	n.d.	n.d.
Zn	μg/L	3000	n.d.	33±14	n.d.	n.d.	nd
Al	μg/L	200	n.d.	n.d.	837±116	511±32	487±50

Table S6.3. Organic characteristics of the surface waters, bank filtrates and groundwater sources

Parameter	Unit	River Nile	Aswan Dam Lake	BF (site1)	BF (site2)	Ambient GW
DOC	mg/L	3.90	3.60	4.30	4.90	5.50
SUVA	L /mg .m	1.56	1.33	2.16	2.04	2.73
HIX	---	0.56	0.55	0.84	0.84	0.86
FIX	---	1.67	1.36	1.09	1.07	1.54
BIX	---	0.71	0.83	0.59	0.53	0.68
P1	R.U.	0.32	0.20	0.41	0.74	1.10
P2	R.U.	0.12	0.09	0.14	0.25	0.46
P3	R.U.	0.23	0.22	0.11	0.14	0.19

7

CONCLUSIONS AND FUTURE PERSPECTIVES

7.1 INTRODUCTION

Access to clean and affordable drinking water is a cross-cutting goal among the United Nations Sustainable Development Goals (SDGs), contributing directly to global development and peace. However, pollution of surface water systems by various contaminants that are difficult to remove using conventional treatment processes impedes the achievement of this goal. Developed countries have relied upon natural treatment techniques, such as BF, for hundreds of years to provide drinking water of adequate quality. In recent years, interest in implementing BF in countries with different hydrological and environmental regimes (e.g., Egypt and India) has increased to meet drinking water demand. However, BF remains a site-specific technique and therefore, there are no standards that can be followed to guarantee the successful implementation of this technique.

The main aims of this study were to assess the reliability of BF in removing chemical pollutants in arid climatic conditions and develop guidelines for the application and sustainable management of BF in Egypt and countries with similar environmental conditions. The research was separated into two phases. Firstly, laboratory-scale batch and column experiments were conducted to assess the impact of different environmental conditions on the removal of chemical pollutants (OM, OMPs, and HMs). An additional series of experiments was then conducted under anaerobic conditions to evaluate the mobilisation of elements (e.g., iron, manganese, and arsenic) from soil and their transfer to bank filtrate. Second, a three-step framework was used to evaluate the effectiveness of BF in arid countries.

7.2 FATE OF ORGANIC MATTER DURING THE BANK FILTRATION PROCESS

DOM is a critical variable that controls the biochemical and redox processes taking place during BF. Moreover, DOM is considered to be a key substrate in carcinogenic disinfection by-product (DBP) formation. A better understanding of the behaviour of DOM during the BF process will help to increase its removal efficiency and optimise the overall performance of BF systems. Laboratory-scale batch and column experiments were performed under different environmental conditions of redox and temperature (20–30 °C) using different water types with different organic compositions. It is recognised that the principal mechanisms of organic matter removal during BF are adsorption and biodegradation. This research highlighted that the removal efficiency of organic matter is strongly regulated by environmental conditions in the infiltration zone (i.e., temperature and the redox environment) with relatively higher removal achieved in an oxic filtration zone and at low temperatures. DOM components were tracked using different analytical techniques (EEM-PARAFAC and size exclusion LC-OCD) to determine the fate of the organic fraction in the infiltration zone.

Experimental results revealed that the organic fraction was composed of biogenic organic compounds (e.g., proteins) that are subjected to strong attenuation during the filtration process. The degradation of these compounds is independent of the infiltrated water temperature ($P > 0.05$). The removal efficiency of biopolymers exceeded 80% at all tested temperatures (20–30 °C). Conversely, the removal of humic compounds was highly dependent on temperature, with higher removal observed at a lower temperature. This difference is primarily attributable to the high capacity of these refractory compounds to adsorb onto sand grains at lower temperatures. However, an increase in the concentration of humic compounds was detected at higher temperatures (30 °C). Such an increase is mainly due to the presence of micro-organisms capable of (i) bio-transformation of labile compounds into more refractory and complex compounds, and (ii) releasing humic compounds from the soil to the bank filtrate. The results of EEM-PARAFAC indicated that the most temperature-dependent fractions are terrestrial humic components, for which sorption characteristics are negatively correlated with temperature. Regarding redox conditions, the column experiments showed that the organic components were less persistent under aerobic conditions. The degradation of biopolymer compounds decreased by 20–24% as the infiltration region shifted from an oxic to a sub-oxic state. Humic compounds also showed poorer removal performance in a sub-oxic environment. Post-treatment may, therefore, be needed when anaerobic filtration conditions are encountered.

7.3 REMOVAL OF ORGANIC MICROPOLLUTANTS DURING BANK FILTRATION

OMPs are of increasing environmental concern when present in surface water systems. These organic compounds are characterised by their variety and toxicity, and are poorly removed by conventional treatment techniques. BF has been acknowledged as a reliable means of eliminating such contaminants to some extent. However, the efficacy of OMP removal is still not fully understood in extreme environments with high temperatures and redox conditions. This research focussed on evaluating the removal efficiencies of various classes of OMPs (herbicides, insecticides, and PAHs) with different physicochemical properties, during BF. Laboratory-scale batch experiments were performed to determine the effect of environmental factors (temperatures in the range 20–30 °C, redox conditions, and raw water composition) on the behaviour of OMPs in the infiltration region.

The experimental results revealed that moderatly hydrophobic OMPs with a solubility (logS) approximately between -2.5 and -4 were most persistent under all of the tested environmental conditions. The removal efficiencies of simazine, atrazine, metolachlor, and isoproturon ranged from 16% to 59% at a residence time of 30 days. The removal efficiencies of these compounds were enhanced at higher temperatures and under aerobic conditions, and when the raw water contained a higher concentration of biodegradable matter. This suggests that co-metabolism and co-adsorption are vital functions in the degradation of these substances during filtration. Furthermore, a lower removal efficiency was achieved when the raw water had a higher concentration of humic compounds, primarily due to the competition between moderate-hydrophobic OMPs and humic compounds for adsorption sites. Conversely, highly hydrophobic compounds (logS < -4) showed relatively high removal efficiency independent of environmental influences. The removal efficiencies of DDT, pyriproxyfen, pendimethalin, β-BHC, endosulfan sulphate, and PAHs exceeded 80%.. Similar removal efficiencies were obtained under abiotic conditions (implemented by the injection of 20 mM of HgCl2 into the systems to inhibit biological activity). This indicates that adsorption is the dominant mechanism in the removal of such compounds during the filtration process. Soluble OMPs with logS higher than -2.5 exhibited the highest degradation capacities during the infiltration process. The removal efficiencies of molinate, propanil, and dimethoate were greater than 70% under the environmental conditions examined. Such removals were, however, limited to < 35% under abiotic conditions, indicating that biodegradation plays a vital function in the attenuation of these compounds during filtration , with preferential removal in the aerobic filtration zone. Moreover, the removal of these OMPs was found to be temperature-dependent, especially when the raw water contained a low concentration of organic matter. However, the extent of this effect reduces at higher temperatures (> 25 °C), which could be attributed to enhanced biological activity.

7.4 REMOVAL OF HEAVY METALS DURING BANK FILTRATION

HMs are categorised as poisonous and non-biodegradable substances that can bioaccumulate in living cells over long periods and which have detrimental effects on human health. Such chemical elements are ubiquitous in natural water environments and are poorly removed by engineered water treatment systems. The primary purpose of this research was to examine the behaviour of HMs in the infiltration zone. Sand column experiments were developed and operated using different types of feed water of varying organic composition to assess the removal of selected HMs (Pb, Cu, Zn, Ni, and Se) during the filtration process. EEM-PARAFAC and fluorescence indices (the humification index, fluorescence index, and biological index) were used to determine the organic characteristics of the feed waters. A further sequence of column experiments was performed to examine the impact of feed water with different types and concentrations of natural organic matter (humic substances, proteins, and humic/protein mixtures) on the removal of HMs. The experiments were performed in a temperature-controlled room at 30 °C.

The experimental results showed that Pb has significant potential to be adsorbed onto sand grains during the filtration process, irrespective of the environmental conditions. Nonetheless, the adsorption capacity of Pb may decrease by up to 20% if the raw water contains a high concentration of humic compounds (> 20 mg-C/L). Conversely, Cu, Zn, and Ni displayed lower removal efficiencies (65–95%) during the filtration process. These removals were significantly different ($P < 0.05$) when the organic content and composition of the feed water varied; higher removal was was observed when the raw water had a limited carbon source. Humic compounds demonstrated a significant potential to suppress the adsorption of HMs mentioned above; negative correlations were detected between the fluorescence indices and the removal efficiencies of the HMs. Humic compounds are negatively charged and can subsequently form adsorbable complexes with metals. Additionally, they can bind with the functional groups on the surfaces of sand grains and reduce the availability of adsorption sites.

The results demonstrate that increasing the concentration of humic substances in the raw water from 5 to 10 mg-C/L could reduce the metal-removal efficiency by 5–25%. Conversely, biodegradable matter (e.g., tyrosine) was found to enhance metal removal, with a removal efficiency above 95% for Cu when the tyrosine concentration in the feed water was 5 mg-C/L. Likewise, the removal of Se was found to be improved when the raw water contained a higher concentration of biodegradable matter. However, it should be noted that increased concentration of biodegradable matter in the raw water might lead to an anaerobic infiltration environment and subsequently, facilitate the transfer of metals from the soil to the infiltrated water.

7.5 MOBILISATION OF IRON, MANGANESE AND ARSENIC DURING BANK FILTRATION

The reductive dissolution of geogenic metals such as Fe and Mn from soil to infiltrated water is one of the main hurdles facing the expansion of the application of BF, especially in arid environments where surface water sources have low dissolved oxygen concentration and therefore, a higher probability of anaerobic conditions developing during infiltration. The release of these metals is also followed by a increase in the concentrations of toxic metals (e.g., As, B, and Cd) in bank filtrate, which greatly degrades its quality and has a severe impact on human health. Several studies have highlighted the role of DOM and its impact on the mobilisation of metals in the infiltration region. During this research, column studies were used to examine the mobilisation of Fe, Mn, and As during BF and the effect of the organic composition of the raw water. The columns were packed with iron-coated sand (1–3 mm) and ripened under anaerobic conditions and fed with water with different organic matter composition. The organic characteristics of the feed waters were defined by coupling F-EEM with the parallel factor framework-clustering analysis tool. Moreover, fluorescence indices were estimated as indicators of the organic composition of the raw water. These experiments were performed in a temperature-controlled room at 30 °C.

The results of these experiments showed that the Mn-reducing mechanism dominated under the tested conditions; the effluent concentration varied from 1,500–3,900 µg/L for Mn, 10–20 µg/L for Fe, and > 2–7.1 µg/L for As. The mobilisation of these metals was found to be significantly impacted by the DOM concentration in the raw water with correlations with the effluent concentrations of Fe, Mn, As of 0.89, 0.87, and 0.63, respectively. Results of the fluorescence indices demonstrated that the mobilisation of metals, particularly Fe and Mn, is highly dependent on the humic content of the raw water. In comparison, humic compounds were found to be less effective at mobilising As. Humic compounds, which have high electron shuttle ability, may bind to metals and consequently improve their mobility. They may also serve as catalysts for the microbial reduction of metals during the filtration process. This research revealed that the concentration of mobilised Mn in the bank filtrate was positively correlated with the humic compound concentrations in the raw water irrespective of their origin. In comparison, mobilised Fe showed a higher affinity to terrestrial humic compounds (which have higher molecular weights).

To validate these findings, a series of batch studies were performed under anaerobic conditions using different types and concentrations of natural organic matter (humic substances, fulvic compounds, and proteins). The experimental results showed that both humic and fulvic compounds (specifically lower-molecular-weight humic compounds) at low concentrations had similar capabilities to release Mn during the filtration process. Nevertheless, fulvic compounds (at concentrations > 5 mg-C/L) had a greater ability to mobilise Mn than humic compounds. Conversely, humic compounds (at concentrations > 10 mg-C/ L) showed a relatively high capacity to mobilise Fe. Biodegradable matter has also been shown to influence the mobilisation of metals in the infiltration zone. A correlation was detected between mobilised Mn and protein-like fluorescent components, demonstrating that microbial reduction plays a role in the release of this element during infiltration.. In contrast, the mobilisation of Fe was influenced to a lesser extent by variations in the concentration of biodegradable DOM in raw water. Nevertheless, batch experiments revealed that the concentration of labile compounds (i.e., tyrosine at > 10 mg-C/L) are adequate to develop a Fe-reducing environment in the infiltration zone and enrich the concentration of Fe in the final pumped water.

154

7.6 ANALYSIS OF THE PERFORMANCE OF BANK FILTRATION IN ARID CLIMATES

The efficacy of BF is widely dependent on the hydrology and the interaction between the surface water system and the adjoining aquifer at the proposed site. Hydrological parameters are highly variable, particularly in hot arid climates. This research was conducted to examine the effectiveness of tBF in the Aswan region of Egypt to determine the degree to which variations in environmental parameters may influence the efficacy of this technique and to provide guidance of the 'best practice' under such environmental conditions. The following three-step methodology was followed to achieve these objectives: (i) develop a hydrological model for the case study; (ii) determine bank filtrate quality; and (iii) examine the economic feasibility of BF in Aswan City, Egypt. A groundwater model was established and validated for a ten-year period to simulate the current hydrological situation in the study region. Then, the numerical model was executed with different environmental conditions to simulate different management scenarios for the BF well fields. Two different potential sites were investigated.

From a hydrological perspective, BF is a promising technique for the Aswan region. However, BF design parameters (e.g., the number of wells, well distance/spacing, distance from the river, and the pumping rate) affect the quality and quantity of the pumped water and thus, must be carefully considered. At the two investigated sites, it was found that situating BF wells very close to the river (< 100 m) may minimise travel time and reduce the treatment efficiency; a low pumping rate could be employed to solve this issue. Moreover, over-abstraction practices could reduce the travel time and induce the flow of contaminated groundwater toward the pumping wells and thus, degrade the quality of bank filtrate. Declining surface water levels also have a substantial effect on the efficiency of BF at the start of service, as it reduces travel time and the share of bank filtrate in the final pumped water. However, the constant operation of wells over a long period could reduce the consequences of this. Samples from surface water sources, groundwater wells, and observation wells drilled at river banks were collected monthly to assess the quality of the bank filtrate over a period of one year. The contribution of river infiltrated water to the total pumped water at the two observation wells was estimated using total conductivity and chloride as cindicator elements. The share of bank filtrate ranged between 70–83% and 30–37% at the two potential sites, respectively

Laboratory results showed that BF is effective in improving the quality of drinking water to Aswan City. The produced water meets the required standards proposed by the WHO (2011). However, organic enrichment of the pumped water was observed. The fluorescence results showed that protein-like compounds were the most visible organic fraction to be eliminated during the filtration process, with removal efficiencies in the range 33–55%, which is in agreement with the findings of laboratory experiments conducted in this study. On the other hand, the concentration of humic compounds was higher in the infiltrated water, which could be attributed to (i) the dissolution of soil organic matter into the infiltrated water and (ii) the mixing of infiltrated water with ambient groundwater with a higher concentration of humic compounds. The enrichment of humic compounds in bank filtrate increases the potential for trihalomethane formation during the post-treatment chlorination process. Thus, a post-treatment process may be required to eliminate such hydrophobic compounds. Ultimately, an economic comparison was made between current treatment techniques and BF using NPV and PBP as indicators. The results revealed that the payback and the NPV of a BF system were lower than for other techniques, which indicates that BF is economically feasible in the case studies examined in this region.

7.7 PRACTICAL IMPLICATIONS AND FURTHER RESEARCH

The findings of this study indicate that BF is an efficient and sustainable way of eliminating chemical contaminants and ensuring good quality drinking water in hot arid countries. However, the efficacy of BF may be optimised through the effective design of the potential well field, taking into account the composition of the raw water and the local hydrological and environmental conditions. The following practical implications are proposed:

Biodegradable organic matter could be adequately eliminated during BF irrespective of the temperature, increasing the biological stability of the bank filtrate. Conversely, humic compounds are likely to be enriched during BF, particularly at higher temperatures. Therefore, post-treatment processes should focus primarily on the elimination of these hydrophobic compounds.

This study suggested using log S (in multiple regression equations) to model the behaviour of OMP compounds in the infiltration region. Moreover, this research indicates the ability of BF to eliminate OMPs in the infiltration zone. However, moderate hydrophobic compounds (-2.5 > log S > -4) have been observed as the most persistent compounds in the infiltration region. The removal of such compounds is enhanced at higher temperatures and under oxic conditions. However, their removal is still insufficient (< 60%) and a long infiltration time is required (residence time = 30 days). If there is a risk of inducing anaerobic conditions, then post-treatment will be reqired. . A combination of BF and membrane-filtration could be an efficient method for eliminating such persistent compounds, but this needs further investigation.

Potential BF sites located in the vicinity of surface water systems with low organic contents (specifically humic compounds) are more feasible to implement BF. Humic compounds reduce the efficiency of BF in eliminating HMs from the infiltrated water.

Moreover, these compounds enhance the mobilisation of metals (e.g., Fe, Mn, and As) from the soil to the bank filtrate and thus, degrade the bank filtrate quality.

The design of BF fields and management plans in Egypt should focus on reducing the percentage of contaminated groundwater in the total pumped water. The consequences that could arise from a decline in surface water levels can be tackled with the continuous operation of BF wells.

Further research is now needed to determine the effect of soil composition on the efficacy of BF in the removal of chemical contaminants. One of the disadvantages limiting the application of this cost-effective technique in arid countries is the high clogging rate in the infiltration zone. As such, more research should be conducted to determine the effects of environmental conditions (e.g., the quality characteristics of raw water) on the clogging process and how this can be minimised.

REFERENCES

Abdel-Fattah, A., Langford, R., & Schulze-Makuch, D. (2008). Applications of particle-tracking techniques to bank infiltration: a case study from El Paso, Texas, USA. *Environmental Geology, 55*(3), 505-515. Retrieved from https://doi.org/10.1007/s00254-007-0996-z. doi:10.1007/s00254-007-0996-z

Abdel-Halim, K. Y., Salama, A. K., El-khateeb, E. N., & Bakry, N. M. (2006). Organophosphorus pollutants (OPP) in aquatic environment at Damietta Governorate, Egypt: Implications for monitoring and biomarker responses. *Chemosphere, 63*(9), 1491-1498. Retrieved from http://www.sciencedirect.com/science/article/pii/S0045653505010969. doi:https://doi.org/10.1016/j.chemosphere.2005.09.019

Abdelhalim, A., Sefelnasr, A., & Ismail, E. (2020). Response of the interaction between surface water and groundwater to climate change and proposed megastructure. *Journal of African Earth Sciences, 162*, 103723. Retrieved from http://www.sciencedirect.com/science/article/pii/S1464343X19303784. doi:https://doi.org/10.1016/j.jafrearsci.2019.103723

Abdelrady, A., Sharma, S., Sefelnasr, A., Abogbal, A., & Kennedy, M. (2019). Investigating the impact of temperature and organic matter on the removal of selected organic micropollutants during bank filtration: A batch study. *Journal of Environmental Chemical Engineering, 7*(1), 102904. Retrieved from http://www.sciencedirect.com/science/article/pii/S2213343719300272. doi:https://doi.org/10.1016/j.jece.2019.102904

Abdelrady, A., Sharma, S., Sefelnasr, A., & Kennedy, M. (2018). The Fate of Dissolved Organic Matter (DOM) During Bank Filtration under Different Environmental Conditions: Batch and Column Studies. *Water, 10*(12), 1730. Retrieved from http://www.mdpi.com/2073-4441/10/12/1730.

Abdelrady, A., Sharma, S., Sefelnasr, A., & Kennedy, M. (2020). Characterisation of the impact of dissolved organic matter on iron, manganese, and arsenic mobilisation during bank filtration. *Journal of Environmental Management, 258*, 110003. Retrieved from http://www.sciencedirect.com/science/article/pii/S0301479719317219. doi:https://doi.org/10.1016/j.jenvman.2019.110003

Abel, C., Sharma, S. K., Maeng, S. K., Magic-Knezev, A., Kennedy, M. D., & Amy, G. L. (2013). Fate of Bulk Organic Matter, Nitrogen, and Pharmaceutically Active Compounds in Batch Experiments Simulating Soil Aquifer Treatment (SAT) Using Primary Effluent. *Water, Air, & Soil Pollution, 224*(7), 1628. Retrieved from https://doi.org/10.1007/s11270-013-1628-8. doi:10.1007/s11270-013-1628-8

Abel, C., Sharma, S. K., Malolo, Y. N., Maeng, S. K., Kennedy, M. D., & Amy, G. L. (2012). Attenuation of Bulk Organic Matter, Nutrients (N and P), and Pathogen Indicators During Soil Passage: Effect of Temperature and Redox Conditions in Simulated Soil Aquifer Treatment (SAT). *Water, Air, & Soil Pollution, 223*(8), 5205-5220. Retrieved from https://doi.org/10.1007/s11270-012-1272-8. doi:10.1007/s11270-012-1272-8

Aboulroos, S., & Satoh, M. (2017). Challenges in Exploiting Resources—General Conclusion. In M. Satoh & S. Aboulroos (Eds.), *Irrigated Agriculture in Egypt: Past, Present and Future* (pp. 267-283). Cham: Springer International Publishing.

Abushaban, A., Mangal, M., Salinas-Rodriguez, S. G., Nnebuo, C., Mondal, S., Goueli, S., . . . Kennedy, M. (2017). *Direct measurement of ATP in seawater and application of ATP to monitor bacterial growth potential in SWRO pre-treatment systems* (Vol. 99).

Ahmed, I. M., Helal, A. A., El Aziz, N. A., Gamal, R., Shaker, N. O., & Helal, A. A. (2015). Influence of some organic ligands on the adsorption of lead by agricultural soil. *Arabian Journal of Chemistry*. Retrieved from http://www.sciencedirect.com/science/article/pii/S1878535215000817. doi:https://doi.org/10.1016/j.arabjc.2015.03.012

Alidina, M., Shewchuk, J., & Drewes, J. E. (2015). Effect of temperature on removal of trace organic chemicals in managed aquifer recharge systems. *Chemosphere, 122*, 23-31. Retrieved from http://www.sciencedirect.com/science/article/pii/S0045653514012491. doi:https://doi.org/10.1016/j.chemosphere.2014.10.064

Anawar, H. M., Akai, J., Komaki, K., Terao, H., Yoshioka, T., Ishizuka, T., . . . Kato, K. (2003). Geochemical occurrence of arsenic in groundwater of Bangladesh: sources and mobilization processes. *Journal of Geochemical Exploration, 77*(2), 109-131. Retrieved from http://www.sciencedirect.com/science/article/pii/S037567420200273X. doi:https://doi.org/10.1016/S0375-6742(02)00273-X

Andersen, C. M., & Bro, R. (2003). Practical aspects of PARAFAC modeling of fluorescence excitation-emission data. *Journal of Chemometrics, 17*(4), 200-215. Retrieved from https://onlinelibrary.wiley.com/doi/abs/10.1002/cem.790. doi:doi:10.1002/cem.790

Anderson, M. P. (2005). Heat as a Ground Water Tracer. *Ground Water, 43*(6), 951-968. Retrieved from http://dx.doi.org/10.1111/j.1745-6584.2005.00052.x. doi:10.1111/j.1745-6584.2005.00052.x

Awad, J., van Leeuwen, J., Chow, C., Drikas, M., Smernik, R. J., Chittleborough, D. J., & Bestland, E. (2016). Characterization of dissolved organic matter for prediction of trihalomethane formation potential in surface and sub-surface waters. *Journal of Hazardous Materials, 308*, 430-439.

Awan, M., Qazi, I., & Khalid, I. (2003). Removal of heavy metals through adsorption using sand. *Journal of environmental sciences (China), 15*, 413-416.

Azadpour-Keeley, A. (2003). *Movement and longevity of viruses in the subsurface* [Cincinnati, OH]: U.S. Environmental Protection Agency, National Risk Management Research Laboratory.

Baczynski, T. P., Pleissner, D., & Grotenhuis, T. (2010). Anaerobic biodegradation of organochlorine pesticides in contaminated soil – Significance of temperature and availability. *Chemosphere, 78*(1), 22-28. doi:http://dx.doi.org/10.1016/j.chemosphere.2009.09.058

Badr, E.-S. A. (2016). Spatio-temporal variability of dissolved organic nitrogen (DON), carbon (DOC), and nutrients in the Nile River, Egypt. *Environmental Monitoring and Assessment, 188*(10), 580. Retrieved from https://doi.org/10.1007/s10661-016-5588-5. doi:10.1007/s10661-016-5588-5

Baghoth, S. A., Sharma, S. K., & Amy, G. L. (2011). Tracking natural organic matter (NOM) in a drinking water treatment plant using fluorescence excitation–emission matrices and PARAFAC. *Water Research, 45*(2), 797-809. Retrieved from http://www.sciencedirect.com/science/article/pii/S0043135410006342. doi:https://doi.org/10.1016/j.watres.2010.09.005

Baker, A. (2001). Fluorescence Excitation–Emission Matrix Characterization of Some Sewage-Impacted Rivers. *Environmental science & technology, 35*(5), 948-953. Retrieved from http://dx.doi.org/10.1021/es000177t. doi:10.1021/es000177t

Barcelona, M. J., & Holm, T. R. (1992). Oxidation-reduction capacities of aquifer solids. [Erratum to document cited in CA115(10):98709d]. *Environmental science & technology,* 26(12), 2540-2540. Retrieved from http://dx.doi.org/10.1021/es00036a033. doi:10.1021/es00036a033

Bartak, R., Grischek, T., Ghodeif, K., & Wahaab, R. (2014). Shortcomings of the RBF Pilot Site in Dishna, Egypt. *Journal of Hydrologic Engineering, 0*(0), 05014033. Retrieved from http://ascelibrary.org/doi/abs/10.1061/%28ASCE%29HE.1943-5584.0001137. doi:10.1061/(ASCE)HE.1943-5584.0001137

Bartak, R., Page, D., Sandhu, C., Grischek, T., Saini, B., Mehrotra, I., . . . Ghosh, N. C. (2015). Application of risk-based assessment and management to riverbank filtration sites in India. *J Water Health, 13*(1), 174-189. Retrieved from https://www.ncbi.nlm.nih.gov/pubmed/25719477. doi:10.2166/wh.2014.075

Berner, R. A. (1981). A new geochemical classification of sedimentary environments. *Journal of Sedimentary Research, 51*(2).

Bertelkamp, C., Reungoat, J., Cornelissen, E., Singhal, N., Reynisson, J., Cabo, A., . . . Verliefde, A. R. (2014). Sorption and biodegradation of organic micropollutants during river bank filtration: A laboratory column study. *Water Research, 52,* 231-241. Retrieved from http://www.sciencedirect.com/science/article/pii/S0043135413008944. doi:http://dx.doi.org/10.1016/j.watres.2013.10.068

Bertelkamp, C., Schoutteten, K., Hulpiau, L., Vanhaecke, L., Vanden Bussche, J., Cabo, A. J., . . . Verliefde, A. (2016a). The effect of feed water dissolved organic carbon concentration and composition on organic micropollutant removal and microbial diversity in soil columns simulating river bank filtration. *Chemosphere, 144,* 932-939. doi:10.1016/j.chemosphere.2015.09.017

Bertelkamp, C., Verliefde, A. R. D., Schoutteten, K., Vanhaecke, L., Vanden Bussche, J., Singhal, N., & van der Hoek, J. P. (2016b). The effect of redox conditions and adaptation time on organic micropollutant removal during river bank filtration: A laboratory-scale column study. *Science of The Total Environment, 544,* 309-318. doi:http://dx.doi.org/10.1016/j.scitotenv.2015.11.035

Beyene, T., Lettenmaier, D. P., & Kabat, P. (2010). Hydrologic impacts of climate change on the Nile River Basin: implications of the 2007 IPCC scenarios. *Climatic Change, 100*(3), 433-461. Retrieved from https://doi.org/10.1007/s10584-009-9693-0. doi:10.1007/s10584-009-9693-0

Bonton, A., Bouchard, C., Barbeau, B., & Jedrzejak, S. (2012). Comparative life cycle assessment of water treatment plants. *Desalination, 284*, 42-54. Retrieved from http://www.sciencedirect.com/science/article/pii/S0011916411007375. doi:https://doi.org/10.1016/j.desal.2011.08.035

Bosuben, N. (2007). *Framework for Feasibility of Bank Filtration Technology for Water Treatment in Developing Countries.* (MSc Thesis), UNESCO-IHE, Delft, The Netherlands

Bourg, A. C., & Bertin, C. (1993). Biogeochemical processes during the infiltration of river water into an alluvial aquifer. *Environmental science & technology, 27*(4), 661-666.

Bourg, A. C. M., & Bertin, C. (1994). Seasonal and Spatial Trends in Manganese Solubility in an Alluvial Aquifer. *Environmental science & technology, 28*(5), 868-876. Retrieved from https://doi.org/10.1021/es00054a018. doi:10.1021/es00054a018

Boving, T. B., Choudri, B. S., Cady, P., Cording, A., Patil, K., & Reddy, V. (2014). Hydraulic and Hydrogeochemical Characteristics of a Riverbank Filtration Site in Rural India. *Water Environment Research, 86*(7), 636-648. doi:10.2175/106143013x13596524516428

Boving, T. B., Patil, K., D'Souza, F., Barker, S. F., McGuinness, S. L., O'Toole, J., . . . Leder, K. (2018). Performance of Riverbank Filtration under Hydrogeologic Conditions along the Upper Krishna River in Southern India. *Water, 11*(1), 12. Retrieved from https://www.mdpi.com/2073-4441/11/1/12.

Brettar, I., Sanchez-Perez, J.-M., & Trémolières, M. (2002). Nitrate elimination by denitrification in hardwood forest soils of the Upper Rhine floodplain – correlation with redox potential and organic matter. *Hydrobiologia, 469*(1-3), 11-21. Retrieved from http://dx.doi.org/10.1023/A%3A1015527611350. doi:10.1023/A:1015527611350

Brune, A., Schink, B., Benz, M., & Kappler, A. (2004). Electron shuttling via humic acids in microbial iron(III) reduction in a freshwater sediment. *FEMS Microbiology Ecology, 47*(1), 85-92. Retrieved from https://dx.doi.org/10.1016/S0168-6496(03)00245-9. doi:10.1016/S0168-6496(03)00245-9 %J FEMS Microbiology Ecology

Calace, N., Deriu, D., Petronio, B. M., & Pietroletti, M. (2009). Adsorption Isotherms and Breakthrough Curves to Study How Humic Acids Influence Heavy Metal–Soil Interactions. *Water, Air, and Soil Pollution, 204*(1), 373-383. Retrieved from https://doi.org/10.1007/s11270-009-0051-7. doi:10.1007/s11270-009-0051-7

Caldwell, T. (2006). Presentation of Data for Factors Significant to Yield from Several Riverbank Filtration Systems in the U.S. and Europe. In S. Hubbs (Ed.), *Riverbank Filtration Hydrology* (Vol. 60, pp. 299-344): Springer Netherlands.

Champ, D. R., Gulens, J., & Jackson, R. E. (1979). Oxidation–reduction sequences in ground water flow systems. *Canadian Journal of Earth Sciences, 16*(1), 12-23. Retrieved from http://cjes.geoscienceworld.org/content/16/1/12.abstract. doi:10.1139/e79-002

Chaweza, D. P. (2006). *Feasibility of riverbank filtration for water treatment in selected cities of Malawi.* (MSc Thesis), UNESCO-IHE, Delft, The Netherlands, Available from http://worldcat.org /z-wcorg/ database.

Chen, F., Peldszus, S., Elhadidy, A. M., Legge, R. L., Van Dyke, M. I., & Huck, P. M. (2016). Kinetics of natural organic matter (NOM) removal during drinking water biofiltration using different NOM characterization approaches. *Water Research, 104,* 361-370. Retrieved from http://www.sciencedirect.com/science/article/pii/S0043135416306285. doi:https://doi.org/10.1016/j.watres.2016.08.028

Cheng, C., Wu, J., You, L., Tang, J., Chai, Y., Liu, B., & Khan, M. F. S. (2018). Novel insights into variation of dissolved organic matter during textile wastewater treatment by fluorescence excitation emission matrix. *Chemical Engineering Journal, 335,* 13-21. Retrieved from http://www.sciencedirect.com/science/article/pii/S1385894717317734. doi:https://doi.org/10.1016/j.cej.2017.10.059

Cheng, D., Yu, J., Wang, T., Chen, W., & Guo, P. (2014). Adsorption Characteristics and Mechanisms of Organochlorine Pesticide DDT on Farmland Soils. *Polish Journal of Environmental Studies, 23*(5).

Chianese, S., Iovino, P., Leone, V., Musmarra, D., & Prisciandaro, M. (2017). Photodegradation of Diclofenac Sodium Salt in Water Solution: Effect of HA, NO3− and TiO2 on Photolysis Performance. *Water, Air, & Soil Pollution, 228*(8), 270. Retrieved from https://doi.org/10.1007/s11270-017-3445-y. doi:10.1007/s11270-017-3445-y

Choudhury, B., Ferraris, S., Ashton, R., Powlson, D., & Whalley, W. (2018). The effect of microbial activity on soil water diffusivity. *European Journal of Soil Science, 69*(3), 407-413.

Chu, K. H. (2010). Fixed bed sorption: Setting the record straight on the Bohart–Adams and Thomas models. *Journal of Hazardous Materials, 177*(1), 1006-1012. Retrieved from http://www.sciencedirect.com/science/article/pii/S030438941000035X. doi:https://doi.org/10.1016/j.jhazmat.2010.01.019

Coble, P. G. (1996). Characterization of marine and terrestrial DOM in seawater using excitation-emission matrix spectroscopy. *Marine Chemistry, 51*(4), 325-346. Retrieved from http://www.sciencedirect.com/science/article/pii/0304420395000623. doi:https://doi.org/10.1016/0304-4203(95)00062-3

Coble, P. G., Lead, J., Baker, A., Reynolds, D. M., & Spencer, R. G. M. (2014). *Aquatic Organic Matter Fluorescence* (P. G. Coble, J. Lead, A. Baker, D. M. Reynolds, & R. G. M. Spencer Eds.). Cambridge: Cambridge University Press.

Colin, A. S., & Rasmus, B. (2008). Characterizing dissolved organic matter fluorescence with parallel factor analysis: a tutorial. *Limnology and Oceanography: Methods, 6*(11), 572-579. Retrieved from https://aslopubs.onlinelibrary.wiley.com/doi/abs/10.4319/lom.2008.6.572. doi:doi:10.4319/lom.2008.6.572

Cressie, N. (1990). The origins of kriging. *Mathematical Geology, 22*(3), 239-252. Retrieved from https://doi.org/10.1007/BF00889887. doi:10.1007/bf00889887

Cuss, C. W., Guéguen, C., Andersson, P., Porcelli, D., Maximov, T., & Kutscher, L. (2016). Advanced Residuals Analysis for Determining the Number of PARAFAC Components in Dissolved Organic Matter. *Applied Spectroscopy, 70*(2), 334-346. Retrieved from http://as.osa.org/abstract.cfm?URI=as-70-2-334.

Dahshan, H., Megahed, A. M., Abd-Elall, A. M. M., Abd-El-Kader, M. A.-G., Nabawy, E., Elbana, M. H. J. J. o. E. H. S., & Engineering. (2016). Monitoring of pesticides water pollution-The Egyptian River Nile. *Journal of Environmental Health Science and Engineering, 14*(1), 15. Retrieved from https://doi.org/10.1186/s40201-016-0259-6. doi:10.1186/s40201-016-0259-6

Dash, R. R., Mehrotra, I., Kumar, P., & Grischek, T. (2008). Lake bank filtration at Nainital, India: water-quality evaluation. *Hydrogeology journal, 16*(6), 1089-1099. Retrieved from http://dx.doi.org/10.1007/s10040-008-0295-0. doi:10.1007/s10040-008-0295-0

Delpla, I., Jung, A. V., Baures, E., Clement, M., & Thomas, O. (2009). Impacts of climate change on surface water quality in relation to drinking water production. *Environment International, 35*(8), 1225-1233. Retrieved from http://www.sciencedirect.com/science/article/pii/S0160412009001494. doi:http://dx.doi.org/10.1016/j.envint.2009.07.001

Derx, J., Andreas, H. F., Matthias Z., Liping P., Jack S., & B., A. P. (2012). Evaluating the effect of temperature induced water viscosity and density fl uctuations on virus and DOC removal during river bank fi ltration – a scenario analysis. *River Systems, 20*(3-4), 169–183.

Deyi, M. (2012). *Development of guidelines for design and performance analysis of riverbank filtration for water treatment.* (MSc Thesis), UNESCO-IHE, Delft, the Netherlands,

Diem, S., Rudolf von Rohr, M., Hering, J. G., Kohler, H.-P. E., Schirmer, M., & von Gunten, U. (2013). NOM degradation during river infiltration: Effects of the climate variables temperature and discharge. *Water Research, 47*(17), 6585-6595. Retrieved from http://www.sciencedirect.com/science/article/pii/S0043135413006684. doi:https://doi.org/10.1016/j.watres.2013.08.028

Dinh, Q. T., Li, Z., Tran, T. A. T., Wang, D., & Liang, D. (2017). Role of organic acids on the bioavailability of selenium in soil: A review. *Chemosphere, 184*, 618-635.

Doussan, C., Poitevin, G., Ledoux, E., & Detay, M. (1997). River bank filtration: modelling of the changes in water chemistry with emphasis on nitrogen species. *Journal of Contaminant Hydrology, 25*(1-2), 129-156. Retrieved from http://dx.doi.org/10.1016/s0169-7722(96)00024-1. doi:10.1016/s0169-7722(96)00024-1

Dragon, K., Górski, J., Kruć, R., Drożdżyński, D., & Grischek, T. (2018). Removal of Natural Organic Matter and Organic Micropollutants during Riverbank Filtration in Krajkowo, Poland. *Water, 10*(10), 1457. Retrieved from https://www.mdpi.com/2073-4441/10/10/1457.

Drewes, J. E., & Summers, R. S. (2003). Natural Organic Matter Removal During Riverbank Filtration: Current Knowledge and Research Needs. In C. Ray, G. Melin, & R. B. Linsky (Eds.), *Riverbank Filtration: Improving Source-Water Quality* (pp. 303-309). Dordrecht: Springer Netherlands.

Duncan, D., Pederson, D. T., Shepherd, T. R., & Can, J. D. (1991). Atrazine Used as a Tracer of Induced Recharge. *Ground Water Monitoring & Remediation, 11*(4), 144-150. Retrieved from http://dx.doi.org/10.1111/j.1745-6592.1991.tb00402.x. doi:10.1111/j.1745-6592.1991.tb00402.x

Duruibe, J. O., Ogwuegbu, M., & Egwurugwu, J. (2007). Heavy metal pollution and human biotoxic effects. *International Journal of physical sciences, 2*(5), 112-118.

Eckert, P., & Irmscher, R. (2006). Over 130 years of experience with Riverbank Filtration in Düsseldorf, Germany. *Journal of Water Supply: Research and Technology - AQUA, 55*(4), 283-291. Retrieved from http://www.scopus.com/inward/record.url?eid=2-s2.0-33745587768&partnerID=40&md5=84c1e522045b6d1cb58015fb3327b634. doi:10.2166/aqua.2006.040

Eckert, P., Lamber, R., & Wagner, C. (2008). The impact of climate change on drinking water supply by riverbank filtration. *Water Science & Technology: Water Supply, 8*(3), 319-324. doi:10.2166/ws.2008.077

El-Nashar, W. Y., & Elyamany, A. H. (2018). Managing risks of the Grand Ethiopian Renaissance Dam on Egypt. *Ain Shams Engineering Journal, 9*(4), 2383-2388. Retrieved from http://www.sciencedirect.com/science/article/pii/S2090447917300837. doi:https://doi.org/10.1016/j.asej.2017.06.004

El-Said, A. G., Badawy, N. A., Abdel-Aal, A. Y., & Garamon, S. E. (2011). Optimization parameters for adsorption and desorption of Zn(II) and Se(IV) using rice husk ash: kinetics and equilibrium. *Ionics, 17*(3), 263-270. Retrieved from https://doi.org/10.1007/s11581-010-0505-3. doi:10.1007/s11581-010-0505-3

El-Zehairy, A. A. M. E. (2014). *Assessment of lake - groundwater interactions : Turawa case, Poland.* (MSc Thesis), University of Twente Faculty of Geo-Information and Earth Observation (ITC), Enschede, The Netherlands. Retrieved from http://www.itc.nl/library/papers_2014/msc/wrem/elzehairy.pdf

ESRI, R. (2011). ArcGIS desktop: release 10. *Environmental Systems Research Institute, CA.*

Fadaei, A., Dehghani, M. H., Nasseri, S., Mahvi, A. H., Rastkari, N., Shayeghi, M. J. B. o. E. C., & Toxicology. (2012). Organophosphorous Pesticides in Surface Water of Iran. *88*(6), 867-869. Retrieved from https://doi.org/10.1007/s00128-012-0568-0. doi:10.1007/s00128-012-0568-0

Farr, T. G., & Kobrick, M. (2000). Shuttle radar topography mission produces a wealth of data. *Eos, Transactions American Geophysical Union, 81*(48), 583-585. Retrieved from https://agupubs.onlinelibrary.wiley.com/doi/abs/10.1029/EO081i048p00583. doi:doi:10.1029/EO081i048p00583

Filter, J., Jekel, M., & Ruhl, A. (2017). Impacts of Accumulated Particulate Organic Matter on Oxygen Consumption and Organic Micro-Pollutant Elimination in Bank Filtration and Soil Aquifer Treatment. *Water, 9*(5), 349. Retrieved from http://www.mdpi.com/2073-4441/9/5/349.

Gabor, R. S., Baker, A., McKnight, D. M., & Miller, M. P. (2014). Fluorescence Indices and Their Interpretation. In A. Baker, D. M. Reynolds, J. Lead, P. G. Coble, & R. G. M. Spencer (Eds.), *Aquatic Organic Matter Fluorescence* (pp. 303-338). Cambridge: Cambridge University Press.

Gandy, C. J., Smith, J. W. N., & Jarvis, A. P. (2007). Attenuation of mining-derived pollutants in the hyporheic zone: A review. *Science of The Total Environment, 373*(2–3), 435-446. Retrieved from http://www.sciencedirect.com/science/article/pii/S0048969706008795. doi:http://dx.doi.org/10.1016/j.scitotenv.2006.11.004

Geriesh, M. H., Balke, K.-D., & El-Rayes, A. E. (2008). Problems of drinking water treatment along Ismailia Canal Province, Egypt. *Journal of Zhejiang University. Science. B, 9*(3), 232-242. Retrieved from http://www.ncbi.nlm.nih.gov/pmc/articles/PMC2266887/. doi:10.1631/jzus.B0710634

Gerlach, M., & Gimbel, R. (1999). Influence of humic substance alteration during soil passage on their treatment behaviour. *Water Science and Technology, 40*(9), 231-239. Retrieved from http://www.sciencedirect.com/science/article/pii/S0273122399006617. doi:https://doi.org/10.1016/S0273-1223(99)00661-7

Ghodeif, K., Grischek, T., Bartak, R., Wahaab, R., & Herlitzius, J. (2016). Potential of river bank filtration (RBF) in Egypt. *Environmental Earth Sciences, 75*(8), 1-13. Retrieved from https://doi.org/10.1007/s12665-016-5454-3. doi:10.1007/s12665-016-5454-3

Gimbel, R., Graham, N. J. D., & Collins, M. R. (2006). *Recent progress in slow sand and alternative biofiltration processes*. London, UK.: IWA Publishing.

Gonçalves-Araujo, R., Granskog, M. A., Bracher, A., Azetsu-Scott, K., Dodd, P. A., & Stedmon, C. A. (2016). Using fluorescent dissolved organic matter to trace and distinguish the origin of Arctic surface waters. *Scientific Reports, 6*, 33978. Retrieved from http://dx.doi.org/10.1038/srep33978. doi:10.1038/srep33978

Graeber, D., Gelbrecht, J., Pusch, M. T., Anlanger, C., & von Schiller, D. (2012). Agriculture has changed the amount and composition of dissolved organic matter in Central European headwater streams. *Science of The Total Environment, 438*, 435-446. Retrieved from http://www.sciencedirect.com/science/article/pii/S0048969712011813. doi:https://doi.org/10.1016/j.scitotenv.2012.08.087

Grischek, T., & Paufler, S. (2017). Prediction of Iron Release during Riverbank Filtration. *9*(5), 317. Retrieved from http://www.mdpi.com/2073-4441/9/5/317.

Grischek, T., Schoenheinz, D., & Ray, C. (2003). Siting and Design Issues for Riverbank Filtration Schemes. In C. Ray, G. Melin, & R. Linsky (Eds.), *Riverbank Filtration* (Vol. 43, pp. 291-302): Springer Netherlands.

Gross-Wittke, A., Gunkel, G., & Hoffmann, A. (2010). Temperature effects on bank filtration: redox conditions and physical-chemical parameters of pore water at Lake Tegel, Berlin, Germany. *Journal of Water and Climate Change, 1*(1), 55-66.

Grünheid, S., Amy, G., & Jekel, M. (2005). Removal of bulk dissolved organic carbon (DOC) and trace organic compounds by bank filtration and artificial recharge. *Water Research, 39*(14), 3219-3228. Retrieved from http://www.sciencedirect.com/science/article/pii/S004313540500285X. doi:http://dx.doi.org/10.1016/j.watres.2005.05.030

Grützmacher, G., Wessel, G., Klitzke, S., & Chorus, I. (2010). Microcystin Elimination During Sediment Contact. *Environmental science & technology, 44*(2), 657-662. Retrieved from https://doi.org/10.1021/es9016816. doi:10.1021/es9016816

Guanxing, H., Jichao, S., Ying, Z., Jingtao, L., Yuxi, Z., Jihong, J., . . . Jili, W. (2011, 6-7 Jan. 2011). *Recent Progress in Research on the Adsorption of Lead in Soil.* Paper presented at the 2011 Third International Conference on Measuring Technology and Mechatronics Automation.

Guigue, J., Mathieu, O., Lévêque, J., Mounier, S., Laffont, R., Maron, P. A., . . . Lucas, Y. (2014). A comparison of extraction procedures for water-extractable organic matter in soils. *European Journal of Soil Science, 65*(4), 520-530. Retrieved from http://dx.doi.org/10.1111/ejss.12156. doi:10.1111/ejss.12156

Hamann, E., Stuyfzand, P. J., Greskowiak, J., Timmer, H., & Massmann, G. (2016). The fate of organic micropollutants during long-term/long-distance river bank filtration. *Science of The Total Environment, 545–546*, 629-640. Retrieved from http://www.sciencedirect.com/science/article/pii/S0048969715312134. doi:http://dx.doi.org/10.1016/j.scitotenv.2015.12.057

Hamdan, A., & Abdel Rady, A. (2011). Vulnerability of the groundwater in the Quaternary aquifer at El Shalal-Kema area, Aswan, Egypt. *Arab Journal of Geoscience, 6*(2), 337-358. doi:10.1007/s12517-011-0363-y

Hamdan, A., Sensoy, M., & Mansour, M. (2013). Evaluating the effectiveness of bank infiltration process in new Aswan City, Egypt. *Arabian Journal of Geosciences, 6*(11), 4155-4165. Retrieved from http://dx.doi.org/10.1007/s12517-012-0682-7. doi:10.1007/s12517-012-0682-7

Hansen, A. M., Kraus, T. E. C., Pellerin, B. A., Fleck, J. A., Downing, B. D., & Bergamaschi, B. A. (2016). Optical properties of dissolved organic matter (DOM): Effects of biological and photolytic degradation. *61*(3), 1015-1032. Retrieved from https://aslopubs.onlinelibrary.wiley.com/doi/abs/10.1002/lno.10270. doi:doi:10.1002/lno.10270

Harbaugh, A. W. (1990). A computer program for calculating subregional water budgets using results from the US Geological Survey modular three-dimensional finite-difference ground-water flow model.

Harshman, R. A., & Lundy, M. E. (1994). PARAFAC: Parallel factor analysis. *Computational Statistics & Data Analysis, 18*(1), 39-72. Retrieved from http://www.sciencedirect.com/science/article/pii/0167947394901325. doi:https://doi.org/10.1016/0167-9473(94)90132-5

Heron, G., Pedersen, J., Tjell, J., & Christensen, T. (1993). Redox Buffering Capacity of Aquifer Sediment. In F. Arendt, G. J. Annokkée, R. Bosman, & W. J. Van Den Brink (Eds.), *Contaminated Soil'93* (Vol. 2, pp. 937-938): Springer Netherlands.

Hiller, E., Jurkovič, L., & Bartal, M. (2008). Effect of temperature on the distribution of polycyclic aromatic hydrocarbons in soil and sediment. *Soil and Water Research, 3*(4), 231-240. Retrieved from http://www.agriculturejournals.cz/uniqueFiles/02871.pdf.

Hinkle, D. E., Wiersma, W., & Jurs, S. G. (1988). Applied statistics for the behavioral sciences.

Hiscock, K. M., & Grischek, T. (2002). Attenuation of groundwater pollution by bank filtration. *Journal of Hydrology, 266*(3–4), 139-144. Retrieved from http://www.sciencedirect.com/science/article/pii/S0022169402001580. doi:http://dx.doi.org/10.1016/S0022-1694(02)00158-0

Hoehn, E., & Cirpka, O. A. (2006). Assessing residence times of hyporheic ground water in two alluvial flood plains of the Southern Alps using water temperature and tracers. *Hydrol. Earth Syst. Sci., 10*(4), 553-563. Retrieved from http://www.hydrol-earth-syst-sci.net/10/553/2006/. doi:10.5194/hess-10-553-2006

Hoehn, E., & Scholtis, A. (2011). Exchange between a river and groundwater, assessed with hydrochemical data. *Hydrol. Earth Syst. Sci., 15*(3), 983-988. Retrieved from https://www.hydrol-earth-syst-sci.net/15/983/2011/. doi:10.5194/hess-15-983-2011

Huang, B. B., Yan, D. H., Wang, H., Cheng, B. F., & Cui, X. H. (2013). Impacts of drought on the quality of surface water of the basin. *Hydrol. Earth Syst. Sci. Discuss., 10*(11), 14463-14493. Retrieved from http://www.hydrol-earth-syst-sci-discuss.net/10/14463/2013/. doi:10.5194/hessd-10-14463-2013

Huber, S. A., Balz, A., Abert, M., & Pronk, W. (2011). Characterisation of aquatic humic and non-humic matter with size-exclusion chromatography – organic carbon detection – organic nitrogen detection (LC-OCD-OND). *Water Research, 45*(2), 879-885. doi:https://doi.org/10.1016/j.watres.2010.09.023

Huguet, A., Vacher, L., Relexans, S., Saubusse, S., Froidefond, J. M., & Parlanti, E. (2009). Properties of fluorescent dissolved organic matter in the Gironde Estuary. *Organic Geochemistry, 40*(6), 706-719. Retrieved from http://www.sciencedirect.com/science/article/pii/S0146638009000655. doi:https://doi.org/10.1016/j.orggeochem.2009.03.002

Hülshoff, I., & Grützmacher, G. (2009). *Analysis of the vulnerability of bank filtration systems to climate change by comparing their effectiveness under varying environmental conditions.* Retrieved from

Hunt, H. (2003). American Experience in Installing Horizontal Collector Wells. In C. Ray, G. Melin, & R. Linsky (Eds.), *Riverbank Filtration* (Vol. 43, pp. 29-34): Springer Netherlands.

Jason, B. F., Eran, H., & M., S. R. G. (2010). Fluorescence spectroscopy opens new windows into dissolved organic matter dynamics in freshwater ecosystems: A review. *Limnology and Oceanography, 55*(6), 2452-2462. Retrieved from https://aslopubs.onlinelibrary.wiley.com/doi/abs/10.4319/lo.2010.55.6.2452. doi:doi:10.4319/lo.2010.55.6.2452

Jørgensen, L., Stedmon, C. A., Kragh, T., Markager, S., Middelboe, M., & Søndergaard, M. (2011). Global trends in the fluorescence characteristics and distribution of marine dissolved organic matter. *Marine Chemistry, 126*(1), 139-148. Retrieved from http://www.sciencedirect.com/science/article/pii/S0304420311000569. doi:https://doi.org/10.1016/j.marchem.2011.05.002

Jumean, F., Pappalardo, L., & Abdo, N. (2010). Removal of Cadmium, Copper, Lead and Nickel from Aqueous Solution by White, Yellow and Red United Arab Emirates Sand. *American journal of environmental sciences.*

Kalakodio, L., Odey, E., & Amenay, A. (2017). *Adsorption and Desorption of Lead (Pb) in Sandy Soil Treated by Various Amendments* (Vol. 07).

Kármán, K., Maloszewski, P., Deák, J., Fórizs, I., & Szabó, C. (2013). Transit time determination for a riverbank filtration system using oxygen isotope data and the lumped-parameter model. *Hydrological Sciences Journal, 59*(6), 1109-1116. Retrieved from http://dx.doi.org/10.1080/02626667.2013.808345. doi:10.1080/02626667.2013.808345

Kedziorek, M. A. M., & Bourg, A. C. M. (2009). Electron trapping capacity of dissolved oxygen and nitrate to evaluate Mn and Fe reductive dissolution in alluvial aquifers during riverbank filtration. *Journal of Hydrology, 365*(1–2), 74-78. Retrieved from http://www.sciencedirect.com/science/article/pii/S0022169408005763. doi:http://dx.doi.org/10.1016/j.jhydrol.2008.11.020

Kedziorek, M. A. M., Geoffriau, S., & Bourg, A. C. M. (2008). Organic Matter and Modeling Redox Reactions during River Bank Filtration in an Alluvial Aquifer of the Lot River, France. *Environmental science & technology, 42*(8), 2793-2798. Retrieved from https://doi.org/10.1021/es702411t. doi:10.1021/es702411t

Kothawala, D. N., Stedmon, C. A., Müller, R. A., Weyhenmeyer, G. A., Köhler, S. J., & Tranvik, L. J. (2014). Controls of dissolved organic matter quality: evidence from a large-scale boreal lake survey. *Global Change Biology, 20*(4), 1101-1114.

Kowalczuk, P., Tilstone, G. H., Zabłocka, M., Röttgers, R., & Thomas, R. (2013). Composition of dissolved organic matter along an Atlantic Meridional Transect from fluorescence spectroscopy and Parallel Factor Analysis. *Marine Chemistry, 157*, 170-184. Retrieved from http://www.sciencedirect.com/science/article/pii/S0304420313001783. doi:https://doi.org/10.1016/j.marchem.2013.10.004

Kpagh, J., Sha'Ato, R., Wuana, R., & Tor-Anyiin, T. (2016). Kinetics of Sorption of Pendimethalin on Soil Samples Obtained from the Banks of Rivers Katsina-Ala and Benue, Central Nigeria. *Journal of Geoscience and Environment Protection, 4*, 37-42. doi:10.4236/gep.2016.41004

Kulkarni, H. V., Mladenov, N., Datta, S., & Chatterjee, D. (2018). Influence of monsoonal recharge on arsenic and dissolved organic matter in the Holocene and Pleistocene aquifers of the Bengal Basin. *Science of The Total Environment, 637-638*, 588-599. Retrieved from http://www.sciencedirect.com/science/article/pii/S0048969718316413. doi:https://doi.org/10.1016/j.scitotenv.2018.05.009

Kulkarni, H. V., Mladenov, N., Johannesson, K. H., & Datta, S. (2017). Contrasting dissolved organic matter quality in groundwater in Holocene and Pleistocene aquifers and implications for influencing arsenic mobility. *Applied Geochemistry, 77*, 194-205. Retrieved from http://www.sciencedirect.com/science/article/pii/S0883292716301081. doi:https://doi.org/10.1016/j.apgeochem.2016.06.002

Lamichhane, S., Bal Krishna, K. C., & Sarukkalige, R. (2016). Polycyclic aromatic hydrocarbons (PAHs) removal by sorption: A review. *Chemosphere, 148*, 336-353. Retrieved from http://www.sciencedirect.com/science/article/pii/S0045653516300352. doi:http://dx.doi.org/10.1016/j.chemosphere.2016.01.036

Lamontagne, S., Taylor, A. R., Batlle-Aguilar, J., Suckow, A., Cook, P. G., Smith, S. D., . . . Stewart, M. K. (2015). River infiltration to a subtropical alluvial aquifer inferred using multiple environmental tracers. *Water Resources Research, 51*(6), 4532-4549. Retrieved from https://www.scopus.com/inward/record.uri?eid=2-s2.0-84937512495&doi=10.1002%2f2014WR015663&partnerID=40&md5=58d5b1e44c8d853d746ca393c66ad6c2. doi:10.1002/2014WR015663

LeChevallier, M. W. (2004). Water treatment and pathogen control process efficiency in achieving safe drinking-water. In K.-K. Au (Ed.). London :: Published on behalf of the World Health Organization by IWA Pub.

Leenheer, J. A., & Croué, J.-P. (2003). Peer Reviewed: Characterizing Aquatic Dissolved Organic Matter. *Environmental science & technology, 37*(1), 18A-26A. Retrieved from http://dx.doi.org/10.1021/es032333c. doi:10.1021/es032333c

Li, D., Alidina, M., & Drewes, J. E. (2014). Role of primary substrate composition on microbial community structure and function and trace organic chemical attenuation in managed aquifer recharge systems. *Applied Microbiology and Biotechnology, 98*(12), 5747-5756. doi:10.1007/s00253-014-5677-8

Li, D., Sharp, J. O., Saikaly, P. E., Ali, S., Alidina, M., Alarawi, M. S., . . . Drewes, J. E. (2012). Dissolved Organic Carbon Influences Microbial Community Composition and Diversity in Managed Aquifer Recharge Systems. *Applied and Environmental Microbiology, 78*(19), 6819. Retrieved from http://aem.asm.org/content/78/19/6819.abstract.

Li, P., Lee, S. H., Lee, S. H., Lee, J.-B., Lee, Y. K., Shin, H.-S., & Hur, J. (2016). Seasonal and storm-driven changes in chemical composition of dissolved organic matter: a case study of a reservoir and its forested tributaries. *Environmental Science and Pollution Research, 23*(24), 24834-24845. Retrieved from https://doi.org/10.1007/s11356-016-7720-z. doi:10.1007/s11356-016-7720-z

Li, Z., Liang, D., Peng, Q., Cui, Z., Huang, J., & Lin, Z. (2017). Interaction between selenium and soil organic matter and its impact on soil selenium bioavailability: A review. *Geoderma, 295*, 69-79. Retrieved from http://www.sciencedirect.com/science/article/pii/S0016706116305018. doi:https://doi.org/10.1016/j.geoderma.2017.02.019

Lim, M.-H., Snyder, S. A., & Sedlak, D. L. (2008). Use of biodegradable dissolved organic carbon (BDOC) to assess the potential for transformation of wastewater-derived contaminants in surface waters. *Water Research, 42*(12), 2943-2952. Retrieved from http://www.sciencedirect.com/science/article/pii/S004313540800119X. doi:http://dx.doi.org/10.1016/j.watres.2008.03.008

Liu, G., Fernandez, A., & Cai, Y. (2011). Complexation of Arsenite with Humic Acid in the Presence of Ferric Iron. *Environmental science & technology, 45*(8), 3210-3216. Retrieved from https://doi.org/10.1021/es102931p. doi:10.1021/es102931p

Lopes, A. R., Danko, A. S., Manaia, C. M., & Nunes, O. C. (2013). Molinate biodegradation in soils: natural attenuation versus bioaugmentation. *Applied Microbiology and Biotechnology, 97*(6), 2691-2700. Retrieved from https://doi.org/10.1007/s00253-012-4096-y. doi:10.1007/s00253-012-4096-y

Lovley, D. R., & Phillips, E. J. (1988). Novel mode of microbial energy metabolism: organic carbon oxidation coupled to dissimilatory reduction of iron or manganese. *Applied and Environmental Microbiology, 54*(6), 1472-1480. Retrieved from https://www.ncbi.nlm.nih.gov/pubmed/16347658

https://www.ncbi.nlm.nih.gov/pmc/PMC202682/.

Lu, F., Chang, C.-H., Lee, D.-J., He, P.-J., Shao, L.-M., & Su, A. (2009). Dissolved organic matter with multi-peak fluorophores in landfill leachate. *Chemosphere, 74*(4), 575-582.

Lynch, S., Batty, L., & Byrne, P. (2014). Environmental Risk of Metal Mining Contaminated River Bank Sediment at Redox-Transitional Zones. *Minerals, 4*(1), 52. Retrieved from http://www.mdpi.com/2075-163X/4/1/52.

Ma, L., & Yates, S. R. (2018). Dissolved organic matter and estrogen interactions regulate estrogen removal in the aqueous environment: A review. *Science of The Total Environment, 640-641*, 529-542.

Macquarrie, K., & Al, T. (2008). *The influence of seasonal temperature variation and other factors on the occurrence of dissolved manganese during river bank filtration.*

Maeng, S. K. (2010). *Multiple Objective Treatment Aspects of Bank Filtration: UNESCO-IHE, Delft, PhD Thesis*: Taylor & Francis.

Maeng, S. K., Ameda, E., Sharma, S. K., Grützmacher, G., & Amy, G. L. (2010). Organic micropollutant removal from wastewater effluent-impacted drinking water sources during bank filtration and artificial recharge. *Water Research, 44*(14), 4003-4014. Retrieved from http://www.sciencedirect.com/science/article/pii/S004313541000223X. doi:http://dx.doi.org/10.1016/j.watres.2010.03.035

Maeng, S. K., & Lee, K.-H. (2019). Riverbank Filtration for the Water Supply on the Nakdong River, South Korea. *Water, 11*(1), 129. Retrieved from https://www.mdpi.com/2073-4441/11/1/129.

Maeng, S. K., Salinas R., A., C. N., & Sharma, S. K. (2013). Removal of Pharmaceuticals by Bank Filtration and Artificial Recharge and Recovery. *62*, 435-451. doi:10.1016/b978-0-444-62657-8.00013-6

Maeng, S. K., Sharma, S. K., Abel, C., Magic-Knezev, A., & Amy, G. L. (2011a). Role of biodegradation in the removal of pharmaceutically active compounds with different bulk organic matter characteristics through managed aquifer recharge: batch and column studies. *Water Research, 45*(16), 4722-4736. Retrieved from https://www.ncbi.nlm.nih.gov/pubmed/21802106. doi:10.1016/j.watres.2011.05.043

Maeng, S. K., Sharma, S. K., Lekkerkerker-Teunissen, K., & Amy, G. L. (2011b). Occurrence and fate of bulk organic matter and pharmaceutically active compounds in managed aquifer recharge: A review. *Water Research, 45*(10), 3015-3033. Retrieved from http://www.sciencedirect.com/science/article/pii/S004313541100073X. doi:https://doi.org/10.1016/j.watres.2011.02.017

Maeng, S. K., Sharma, S. K., Magic-Knezev, A., & Amy, G. (2008). Fate of effluent organic matter (EfOM) and natural organic matter (NOM) through riverbank filtration. *Water Sci Technol, 57*(12), 1999-2007. Retrieved from https://www.ncbi.nlm.nih.gov/pubmed/18587190. doi:10.2166/wst.2008.613

Mal'tseva, E. V., & Yudina, N. V. (2014). Sorption of humic acids by quartz sands. *Solid Fuel Chemistry, 48*(4), 239-244. Retrieved from https://doi.org/10.3103/S0361521914040089. doi:10.3103/s0361521914040089

Mansour, S. A., & Sidky, M. M. (2003). Ecotoxicological Studies. 6. The first comparative study between Lake Qarun and Wadi El-Rayan wetland (Egypt), with respect to contamination of their major components. *Food Chemistry, 82*(2), 181-189. doi:https://doi.org/10.1016/S0308-8146(02)00451-X

Martin, S., & Griswold, W. (2009). Human health effects of heavy metals. *Environmental Science and Technology briefs for citizens, 15*, 1-6.

Massmann, G., Greskowiak, J., Dünnbier, U., Zuehlke, S., Knappe, A., & Pekdeger, A. (2006). The impact of variable temperatures on the redox conditions and the behaviour of pharmaceutical residues during artificial recharge. *Journal of Hydrology, 328*(1-2), 141-156. doi:10.1016/j.jhydrol.2005.12.009

Massmann, G., Nogeitzig, A., Taute, T., & Pekdeger, A. (2008). Seasonal and spatial distribution of redox zones during lake bank filtration in Berlin, Germany. *Environmental Geology, 54*(1), 53-65. Retrieved from http://dx.doi.org/10.1007/s00254-007-0792-9. doi:10.1007/s00254-007-0792-9

Matilainen, A., & Sillanpää, M. (2010). Removal of natural organic matter from drinking water by advanced oxidation processes. *Chemosphere, 80*(4), 351-365. Retrieved from http://www.sciencedirect.com/science/article/pii/S0045653510005163. doi:http://dx.doi.org/10.1016/j.chemosphere.2010.04.067

Matsunaga, T., Karametaxas, G., Von Gunten, H., & Lichtner, P. J. G. e. C. A. (1993). Redox chemistry of iron and manganese minerals in river-recharged aquifers: A model interpretation of a column experiment. *57*(8), 1691-1704.

McDonald, M. G., & Harbaugh, A. W. (1988). *A modular three-dimensional finite-difference ground-water flow model* (06-A1). Retrieved from http://pubs.er.usgs.gov/publication/twri06A1

McKnight, D. M., Boyer, E. W., Westerhoff, P. K., Doran, P. T., Kulbe, T., Andersen, D. T. J. L., & Oceanography. (2001). Spectrofluorometric characterization of dissolved organic matter for indication of precursor organic material and aromaticity. *46*(1), 38-48.

McMahon, P. B., & Chapelle, F. H. (2008). Redox Processes and Water Quality of Selected Principal Aquifer Systems. *Ground Water, 46*(2), 259-271. doi:10.1111/j.1745-6584.2007.00385.x

Meffe, R., & de Bustamante, I. (2014). Emerging organic contaminants in surface water and groundwater: A first overview of the situation in Italy. *Science of The Total Environment, 481*, 280-295. Retrieved from http://www.sciencedirect.com/science/article/pii/S0048969714002277. doi:https://doi.org/10.1016/j.scitotenv.2014.02.053

Miller, M. P., McKnight, D. M., Cory, R. M., Williams, M. W., & Runkel, R. L. (2006). Hyporheic Exchange and Fulvic Acid Redox Reactions in an Alpine Stream/Wetland Ecosystem, Colorado Front Range. *Environmental science & technology, 40*(19), 5943-5949. Retrieved from https://doi.org/10.1021/es060635j. doi:10.1021/es060635j

Misahah al-Jiyulujiyah, a.-M., & Attia, M. I. (1954). *Deposits in the Nile Valley and the delta*. Cairo: Government Press.

Misra, A. K. (2014). Climate change and challenges of water and food security. *International Journal of Sustainable Built Environment, 3*(1), 153-165. Retrieved from http://www.sciencedirect.com/science/article/pii/S221260901400020X. doi:https://doi.org/10.1016/j.ijsbe.2014.04.006

Murphy, K. R., Hambly, A., Singh, S., Henderson, R. K., Baker, A., Stuetz, R., & Khan, S. J. (2011). Organic Matter Fluorescence in Municipal Water Recycling Schemes: Toward a Unified PARAFAC Model. *Environmental science & technology, 45*(7), 2909-2916. Retrieved from https://doi.org/10.1021/es103015e. doi:10.1021/es103015e

Murphy, K. R., Stedmon, C. A., Graeber, D., & Bro, R. (2013). Fluorescence spectroscopy and multi-way techniques. PARAFAC. *Analytical Methods, 5*(23), 6557-6566. Retrieved from http://dx.doi.org/10.1039/C3AY41160E. doi:10.1039/C3AY41160E

Murphy, K. R., Stedmon, C. A., Wenig, P., & Bro, R. (2014). OpenFluor– an online spectral library of auto-fluorescence by organic compounds in the environment. *Analytical Methods, 6*(3), 658-661. Retrieved from http://dx.doi.org/10.1039/C3AY41935E. doi:10.1039/C3AY41935E

Nagy-Kovács, Z., Davidesz, J., Czihat-Mártonné, K., Till, G., Fleit, E., & Grischek, T. (2019). Water Quality Changes during Riverbank Filtration in Budapest, Hungary. *Water, 11*(2), 302. Retrieved from https://www.mdpi.com/2073-4441/11/2/302.

Nam, S.-N., & Amy, G. (2008). *Differentiation of wastewater effluent organic matter (EfOM) from natural organic matter (NOM) using multiple analytical techniques* (Vol. 57).

Neidhardt, H., Berner, Z. A., Freikowski, D., Biswas, A., Majumder, S., Winter, J., . . . Norra, S. (2014). Organic carbon induced mobilization of iron and manganese in a West Bengal aquifer and the muted response of groundwater arsenic concentrations. *Chemical Geology, 367*, 51-62. Retrieved from http://www.sciencedirect.com/science/article/pii/S0009254114000138. doi:https://doi.org/10.1016/j.chemgeo.2013.12.021

Nunes, O. C., Lopes, A. R., & Manaia, C. M. (2013). Microbial degradation of the herbicide molinate by defined cultures and in the environment. *Applied Microbiology and Biotechnology, 97*(24), 10275-10291. Retrieved from https://doi.org/10.1007/s00253-013-5316-9. doi:10.1007/s00253-013-5316-9

Ohno, T. (2002). Fluorescence Inner-Filtering Correction for Determining the Humification Index of Dissolved Organic Matter. *Environmental science & technology, 36*(4), 742-746. Retrieved from http://dx.doi.org/10.1021/es0155276. doi:10.1021/es0155276

Orlandini, E. (1999). *Pesticide removal by combined ozonation and granular activated carbon filtration.* (PhD thesis), IHE-UNESCO, Delft. Retrieved from http://edepot.wur.nl/192726

Osburn, C. L., Boyd, T. J., Montgomery, M. T., Bianchi, T. S., Coffin, R. B., & Paerl, H. W. (2016a). Optical Proxies for Terrestrial Dissolved Organic Matter in Estuaries and Coastal Waters. *Frontiers in Marine Science, 2*(127). Retrieved from https://www.frontiersin.org/article/10.3389/fmars.2015.00127. doi:10.3389/fmars.2015.00127

Osburn, C. L., Handsel, L. T., Peierls, B. L., & Paerl, H. W. (2016b). Predicting Sources of Dissolved Organic Nitrogen to an Estuary from an Agro-Urban Coastal Watershed. *Environmental science & technology, 50*(16), 8473-8484. Retrieved from https://doi.org/10.1021/acs.est.6b00053. doi:10.1021/acs.est.6b00053

Pan, W., Huang, Q., & Huang, G. (2018). Nitrogen and Organics Removal during Riverbank Filtration along a Reclaimed Water Restored River in Beijing, China. *Water, 10*(4), 491. Retrieved from https://www.mdpi.com/2073-4441/10/4/491.

Park, M., & Snyder, S. A. (2018). Sample handling and data processing for fluorescent excitation-emission matrix (EEM) of dissolved organic matter (DOM). *Chemosphere, 193*, 530-537. Retrieved from http://www.sciencedirect.com/science/article/pii/S0045653517318465. doi:https://doi.org/10.1016/j.chemosphere.2017.11.069

Paufler, S., Grischek, T., Benso, M. R., Seidel, N., & Fischer, T. (2018). The Impact of River Discharge and Water Temperature on Manganese Release from the Riverbed during Riverbank Filtration: A Case Study from Dresden, Germany. *10*(10), 1476. Retrieved from http://www.mdpi.com/2073-4441/10/10/1476.

Poggenburg, C., Mikutta, R., Schippers, A., Dohrmann, R., & Guggenberger, G. (2018). Impact of natural organic matter coatings on the microbial reduction of iron oxides. *Geochimica et Cosmochimica Acta, 224*, 223-248. Retrieved from http://www.sciencedirect.com/science/article/pii/S0016703718300073. doi:https://doi.org/10.1016/j.gca.2018.01.004

Pollock, D. W. (1989). *Documentation of computer programs to compute and display pathlines using results from the U.S. Geological Survey modular three-dimensional finite-difference ground-water flow model* (89-381). Retrieved from http://pubs.er.usgs.gov/publication/ofr89381

Qian, C., Wang, L.-F., Chen, W., Wang, Y.-S., Liu, X.-Y., Jiang, H., & Yu, H.-Q. (2017). Fluorescence Approach for the Determination of Fluorescent Dissolved Organic Matter. *Analytical Chemistry, 89*(7), 4264-4271. Retrieved from https://doi.org/10.1021/acs.analchem.7b00324. doi:10.1021/acs.analchem.7b00324

Rani, S., & Sud Sant, D. (2014). Time and temperature dependent sorption behaviour of dimethoate pesticide in various indian soils. *International Agrophysics, 28*(4), 479-490. Retrieved from https://www.degruyter.com/view/j/intag.2014.28.issue-4/intag-2014-0038/intag-2014-0038.xml. doi:10.2478/intag-2014-0038

Ray, C., Melin, G., & Linsky, R. B. (2002). *Riverbank filtration : improving source-water quality*: Dordrecht ; Boston : Kluwer Academic Publishers ; Fountain Valley, Calif. : In collaboration with NWRI, National Water Research Institute, c2002.

Ray, C., & Shamrukh, M. (2011). Riverbank Filtration for Water Security in Desert Countries. *NATO Science for Peace and Security Series C: Environmental Security.*

Refaey, Y., Jansen, B., De Voogt, P., Parsons, J. R., El-Shater, A., El-Haddad, A., & Kalbitz, K. (2017a). Influence of Organo-Metal Interactions on Regeneration of Exhausted Clay Mineral Sorbents in Soil Columns Loaded with Heavy Metals.

Pedosphere, 27(3), 579-587. doi:https://doi.org/10.1016/S1002-0160(17)60353-9

Refaey, Y., Jansen, B., Parsons, J. R., De Voogt, P., Bagnis, S., Markus, A., . . . Kalbitz, K. (2017b). Effects of clay minerals, hydroxides, and timing of dissolved organic matter addition on the competitive sorption of copper, nickel, and zinc: A column experiment. *Journal of Environmental Management, 187*, 273-285. Retrieved from http://www.sciencedirect.com/science/article/pii/S0301479716309483. doi:https://doi.org/10.1016/j.jenvman.2016.11.056

Rehman, Z., Jeong, S., Tabatabai, S., Emwas, A., & Leiknes, T. (2017). Advanced characterization of dissolved organic matter released by bloom-forming marine algae. *Desalination and Water Treatment, 69*, 1-11. doi:10.5004/dwt.2017.0444

Reuter, J. H., & Perdue, E. M. (1977). Importance of heavy metal-organic matter interactions in natural waters. *Geochimica et Cosmochimica Acta, 41*(2), 325-334. Retrieved from http://www.sciencedirect.com/science/article/pii/001670377790240X. doi:https://doi.org/10.1016/0016-7037(77)90240-X

Rodríguez-Liébana, J. A., ElGouzi, S., & Peña, A. (2017). Laboratory persistence in soil of thiacloprid, pendimethalin and fenarimol incubated with treated wastewater and dissolved organic matter solutions. Contribution of soil biota. *Chemosphere, 181*, 508-517. Retrieved from http://www.sciencedirect.com/science/article/pii/S0045653517306483. doi:http://dx.doi.org/10.1016/j.chemosphere.2017.04.111

Rohr, M. R. v. (2014). *Effects of climate change on redox processes during riverbank filtration: Field studies and column experiments.* (PhD Thesis), ETH-Zürich, Switzerland,

Romero-Esquivel, L. G., Grischek, T., Pizzolatti, B. S., Mondardo, R. I., & Sens, M. L. (2017). Bank filtration in a coastal lake in South Brazil: water quality, natural organic matter (NOM) and redox conditions study. *Clean Technologies and Environmental Policy, 19*(8), 2007-2020. Retrieved from https://doi.org/10.1007/s10098-017-1382-5. doi:10.1007/s10098-017-1382-5

Rostad, C. E., Leenheer, J. A., Katz, B., Martin, B. S., & Noyes, T. I. (2000). Characterization and Disinfection By-Product Formation Potential of Natural Organic Matter in Surface and Ground Waters from Northern Florida. *761*, 154-172. doi:10.1021/bk-2000-0761.ch011

Sandhu, C., Grischek, T., Börnick, H., Feller, J., & Sharma, S. K. (2019). A Water Quality Appraisal of Some Existing and Potential Riverbank Filtration Sites in India. *Water, 11*(2), 215. Retrieved from https://www.mdpi.com/2073-4441/11/2/215.

Sandhu, C. S. S. (2015). *A Concept for the Investigation of Riverbank Filtration Sites for Potable Water Supply in India.*

Sanyal, D., Yaduraju, N. T., & Kulshrestha, G. (2000). Metolachlor persistence in laboratory and field soils under Indian tropical conditions. *Journal of Environmental Science and Health, Part B, 35*(5), 571-583.. doi:10.1080/03601230009373293

Sastre, J., Sahuquillo, A., Vidal, M., & Rauret, G. (2002). Determination of Cd, Cu, Pb and Zn in environmental samples: microwave-assisted total digestion versus aqua regia and nitric acid extraction. *Analytica Chimica Acta, 462*(1), 59-72. Retrieved from http://www.sciencedirect.com/science/article/pii/S0003267002003070. doi:https://doi.org/10.1016/S0003-2670(02)00307-0

Schijven, J., Berger, P., & Miettinen, I. (2003). Removal of Pathogens, Surrogates, Indicators, and Toxins Using Riverbank Filtration. In C. Ray, G. Melin, & R. Linsky (Eds.), *Riverbank Filtration* (Vol. 43, pp. 73-116): Springer Netherlands.

Schittich, A.-R., Wünsch, U. J., Kulkarni, H. V., Battistel, M., Bregnhøj, H., Stedmon, C. A., & McKnight, U. S. (2018). Investigating Fluorescent Organic-Matter Composition as a Key Predictor for Arsenic Mobility in Groundwater Aquifers. *Environmental science & technology, 52*(22), 13027-13036. Retrieved from https://doi.org/10.1021/acs.est.8b04070. doi:10.1021/acs.est.8b04070

Schmidt, C. K., Lange, F. T., Brauch, H. J., & Kühn, W. (2003). *Experiences with bank filtration and infiltration in Germany.* Retrieved from

Schoenheinz, D. (2004). *DOC as control parameter for the evaluation and management of aquifers with anthropogenic influenced infiltration.* PhD thesis, Faculty of Forestry, Geo and Hydro Sciences, Dresden University …,

Schwarzenbach, R. P., Escher Bi Fau - Fenner, K., Fenner K Fau - Hofstetter, T. B., Hofstetter Tb Fau - Johnson, C. A., Johnson Ca Fau - von Gunten, U., von Gunten U Fau - Wehrli, B., & Wehrli, B. (2006). The challenge of micropollutants in aquatic systems. *Science, 313* (1095-9203 (Electronic)), 1072-1077. doi:10.1126/science.1127291

Seitzinger, S., Harrison, J. A., Böhlke, J., Bouwman, A., Lowrance, R., Peterson, B., . . . Drecht, G. V. (2006). Denitrification across landscapes and waterscapes: a synthesis. *Ecological Applications, 16*(6), 2064-2090.

Selim, M. I., & Popendorf, W. J. (2009). Pesticide Contamination of Surface Water in Egypt and Potential Impact. *CATRINA-THE INTERNATIONAL JOURNAL OF ENVIRONMENTAL SCIENCES, 4*(1), 1-9.

Selim, S., Hamdan, A., & Rady, A. (2014). Groundwater Rising as Environmental Problem, Causes and Solutions: Case Study from Aswan City, Upper Egypt. *Open Journal of Geology, 4*, 324-341. doi:10.4236/ojg.2014.47025

Seybold, C. A., Mersie, W., & McNamee, C. (2001). Anaerobic Degradation of Atrazine and Metolachlor and Metabolite Formation in Wetland Soil and Water Microcosms. *JOURNAL OF ENVIRONMENTAL QUALITY, 30*(4), 1271-1277. Retrieved from http://dx.doi.org/10.2134/jeq2001.3041271x. doi:10.2134/jeq2001.3041271x

Shahgholi, H., & Gholamalizadeh Ahangar, A. (2014). Factors controlling degradation of pesticides in the soil environment: A Review. *Agriculture Science Developments, 3*(8), 273-278.

Shamrukh, M., & Abdel-Wahab, A. (2008). Riverbank filtration for sustainable water supply: application to a large-scale facility on the Nile River. *Clean Technologies and Environmental Policy, 10*(4), 351-358.

Shamrukh, M., & Abdel-Wahab, A. (2011). Water Pollution and Riverbank Filtration for Water Supply Along River Nile, Egypt. In M. Shamrukh (Ed.), *Riverbank Filtration for Water Security in Desert Countries* (pp. 5-28): Springer Netherlands.

Shamrukh, M., Corapcioglu, M., & A.A. Hassona, F. (2005). *Modeling the Effect of Chemical Fertilizers on Ground Water Quality in the Nile Valley Aquifer, Egypt* (Vol. 39).

Sharma, S. K. (2014). *Heavy metals in water: presence, removal and safety*: Royal Society of Chemistry.

Sharma, S. K., & Amy, G. (2009). *Bank filtration: A sustainable water treatment technology for developing countries*. Paper presented at the 34th WEDC International Conference, Addis Ababa, Ethiopia.

Sharma, S. K., Chaweza, D., Bosuben, N., Holzbecher, E., & Amy, G. (2012). Framework for feasibility assessment and performance analysis of riverbank filtration systems for water treatment. *Journal of Water Supply: Research and Technology-Aqua, 61*(2), 73-81.

Shi, Y., Huang, J., Zeng, G., Gu, Y., Hu, Y., Tang, B., . . . Shi, L. (2018). Evaluation of soluble microbial products (SMP) on membrane fouling in membrane bioreactors (MBRs) at the fractional and overall level: a review. *Reviews in Environmental Science and Bio/Technology, 17*(1), 71-85. Retrieved from https://doi.org/10.1007/s11157-017-9455-9. doi:10.1007/s11157-017-9455-9

Shutova, Y., Baker, A., Bridgeman, J., & Henderson, R. K. (2014). Spectroscopic characterisation of dissolved organic matter changes in drinking water treatment: From PARAFAC analysis to online monitoring wavelengths. *Water Research, 54*, 159-169. Retrieved from http://www.sciencedirect.com/science/article/pii/S0043135414000931. doi:https://doi.org/10.1016/j.watres.2014.01.053

Singh, B. K., & Walker, A. (2006). Microbial degradation of organophosphorus compounds. *FEMS Microbiology Reviews, 30*(3), 428-471. Retrieved from http://dx.doi.org/10.1111/j.1574-6976.2006.00018.x. doi:10.1111/j.1574-6976.2006.00018.x

Singh, S., Inamdar, S., & Scott, D. (2013). Comparison of Two PARAFAC Models of Dissolved Organic Matter Fluorescence for a Mid-Atlantic Forested Watershed in the USA. *Journal of Ecosystems, 2013*, 16. Retrieved from http://dx.doi.org/10.1155/2013/532424. doi:10.1155/2013/532424

So, S. H., Choi, I. H., Kim, H. C., & Maeng, S. K. (2017). Seasonally related effects on natural organic matter characteristics from source to tap in Korea. *Science of The Total Environment, 592*, 584-592. Retrieved from http://www.sciencedirect.com/science/article/pii/S0048969717305752. doi:https://doi.org/10.1016/j.scitotenv.2017.03.063

Sontheimer, H. (1980). Experience With Riverbank Filtration Along the Rhine River. *American Water Works Association, 72*(7), 386-390. Retrieved from http://www.jstor.org/stable/41270526. doi:10.2307/41270526

Sophocleous, M. (1997). Managing water resources systems: why "safe yield" is not sustainable. *Groundwater, 35*(4), 561-561

Sophocleous, M. (2000). From safe yield to sustainable development of water resources—the Kansas experience. *Journal of Hydrology, 235*(1), 27-43

Sophocleous, M. (2002). Interactions between groundwater and surface water: the state of the science. *Hydrogeology journal, 10*(1), 52-67

Sprenger, C., Lorenzen, G., & Asaf, P. (2012). Environmental Tracer Application and Purification Capacity at a Riverbank Filtration Well in Delhi (India). *Journal of Indian Water Works Association, Special Issue on River Bank Filtration, 1*, 25-32.

Sprenger, C., Lorenzen, G., Hülshoff, I., Grützmacher, G., Ronghang, M., & Pekdeger, A. (2011). Vulnerability of bank filtration systems to climate change. *Science of The Total Environment, 409*(4), 655-663. Retrieved from http://www.sciencedirect.com/science/article/pii/S0048969710012088. doi:http://dx.doi.org/10.1016/j.scitotenv.2010.11.002

Stahlschmidt, M., Regnery, J., Campbell, A., & Drewes, J. (2015). Application of 3D-fluorescence/PARAFAC to monitor the performance of managed aquifer recharge facilities. *Journal of Water Reuse and Desalination, 6*(2), 249-263. Retrieved from http://dx.doi.org/10.2166/wrd.2015.220. doi:10.2166/wrd.2015.220

Stedmon, C. A., Markager, S., & Bro, R. (2003). Tracing dissolved organic matter in aquatic environments using a new approach to fluorescence spectroscopy. *Marine Chemistry, 82*(3), 239-254. Retrieved from http://www.sciencedirect.com/science/article/pii/S0304420303000720. doi:https://doi.org/10.1016/S0304-4203(03)00072-0

Stedmon, C. A., Markager, S., Tranvik, L., Kronberg, L., Slätis, T., & Martinsen, W. (2007). Photochemical production of ammonium and transformation of dissolved organic matter in the Baltic Sea. *Marine Chemistry, 104*(3), 227-240. Retrieved from http://www.sciencedirect.com/science/article/pii/S0304420306001915. doi:https://doi.org/10.1016/j.marchem.2006.11.005

Stefaniak, J., Dutta, A., Verbinnen, B., Shakya, M., & Rene, E. R. (2018). Selenium removal from mining and process wastewater: a systematic review of available technologies. *Journal of Water Supply: Research and Technology-Aqua, 67*(8), 903-918. Retrieved from https://doi.org/10.2166/aqua.2018.109. doi:10.2166/aqua.2018.109

Stocker, T. F., Qin, D., Plattner, G.-K., Tignor, M., Allen, S. K., Boschung, J., . . . Midgley, P. M. (2013). Climate change 2013: The physical science basis. *Contribution of working group I to the fifth assessment report of the intergovernmental panel on climate change, 1535*.

Sugiyama, S., & Hama, T. (2013). Effects of water temperature on phosphate adsorption onto sediments in an agricultural drainage canal in a paddy-field district. *Ecological Engineering, 61, Part A*(0), 94-99. Retrieved from http://www.sciencedirect.com/science/article/pii/S0925857413003996. doi:http://dx.doi.org/10.1016/j.ecoleng.2013.09.053

Sullivan, J., & Sacramentogoh, K. (2008). Environmental fate and properties of pyriproxyfen. *Journal of Pesticide Science, 33*(4), 339-350. doi:0.1584/jpestics.R08-02

Sun, H. Y., Koal, P., Gerl, G., Schroll, R., Joergensen, R. G., & Munch, J. C. (2017). Water-extractable organic matter and its fluorescence fractions in response to minimum tillage and organic farming in a Cambisol. *Chemical and Biological Technologies in Agriculture, 4*(1), 15. Retrieved from https://doi.org/10.1186/s40538-017-0097-5. doi:10.1186/s40538-017-0097-5

Tian, Y., Sun, W., Qu, D., & Wang, L. (2011, 25-28 March 2011). *Study of a Exhaust After-Treatment System Applied to Hybrid Vehicle.* Paper presented at the 2011 Asia-Pacific Power and Energy Engineering Conference.

Tong, K., Lin, A., Ji, G., Wang, D., & Wang, X. (2016). The effects of adsorbing organic pollutants from super heavy oil wastewater by lignite activated coke. *Journal of Hazardous Materials, 308*, 113-119. Retrieved from http://www.sciencedirect.com/science/article/pii/S0304389416300140. doi:https://doi.org/10.1016/j.jhazmat.2016.01.014

Tran, N. H., Urase, T., Ngo, H. H., Hu, J., & Ong, S. L. (2013). Insight into metabolic and cometabolic activities of autotrophic and heterotrophic microorganisms in the biodegradation of emerging trace organic contaminants. *Bioresource Technology, 146*, 721-731. Retrieved from http://www.sciencedirect.com/science/article/pii/S0960852413011516. doi:http://dx.doi.org/10.1016/j.biortech.2013.07.083

Tufenkji, N., Ryan, J. N., & Elimelech, M. (2002). The Promise of Bank Filtration. *Environmental science & technology, 36*(21), 422A-428A. Retrieved from http://dx.doi.org/10.1021/es022441j. doi:10.1021/es022441j

Van Scoy, A., Pennell, A., & Zhang, X. (2016). Environmental Fate and Toxicology of Dimethoate. In W. P. de Voogt (Ed.), *Reviews of Environmental Contamination and Toxicology Volume 237* (pp. 53-70). Cham: Springer International Publishing.

Van Vliet, M. T. H., & Zwolsman, J. J. G. (2008). Impact of summer droughts on the water quality of the Meuse river. *Journal of Hydrology, 353*(1–2), 1-17. Retrieved from http://www.sciencedirect.com/science/article/pii/S0022169408000024. doi:http://dx.doi.org/10.1016/j.jhydrol.2008.01.001

Vasyukova, E., Proft, R., & Uhl, W. (2014). Evaluation of dissolved organic matter fractions removal due to biodegradation. *Progress in Slow Sand and Alternative Biofiltration Processes: Further Developments and Applications*, 59-66.

Vega, M. A., Kulkarni, H. V., Mladenov, N., Johannesson, K., Hettiarachchi, G. M., Bhattacharya, P., . . . Datta, S. (2017). Biogeochemical Controls on the Release and Accumulation of Mn and As in Shallow Aquifers, West Bengal, India. *Frontiers in Environmental Science, 5*, 29. Retrieved from https://www.frontiersin.org/article/10.3389/fenvs.2017.00029.

Verstraeten, I. M., Soenksen, P. J., Engel, G. B., & Miller, L. D. (1999). Organic compounds in the environment: Determining travel time and stream mixing using tracers and empirical equations. *JOURNAL OF ENVIRONMENTAL QUALITY, 28*(5), 1387-1395. Retrieved from http://pubs.er.usgs.gov/publication/70021980.

Verweij, W., Wiele, J. v. d., Moorselaar, I. v., & Grinten, E. v. d. (2010). *Impact of climate change on water quality in the Netherlands*. Retrieved from

Vogt, T., Hoehn, E., Schneider, P., Freund, A., Schirmer, M., & Cirpka, O. A. (2010). Fluctuations of electrical conductivity as a natural tracer for bank filtration in a losing stream. *Advances in Water Resources, 33*(11), 1296-1308. Retrieved from http://www.sciencedirect.com/science/article/pii/S0309170810000394. doi:http://dx.doi.org/10.1016/j.advwatres.2010.02.007

Wahaab, R. A., Salah, A., & Grischek, T. (2019). Water Quality Changes during the Initial Operating Phase of Riverbank Filtration Sites in Upper Egypt. *Water, 11*(6), 1258. Retrieved from https://www.mdpi.com/2073-4441/11/6/1258.

Walker, S. A., Amon, R. M. W., Stedmon, C., Duan, S., & Louchouarn, P. (2009). The use of PARAFAC modeling to trace terrestrial dissolved organic matter and fingerprint water masses in coastal Canadian Arctic surface waters. *Journal of Geophysical Research: Biogeosciences, 114*(G4). Retrieved from https://agupubs.onlinelibrary.wiley.com/doi/abs/10.1029/2009JG000990. doi:doi:10.1029/2009JG000990

Wang, B., Jin, M., Nimmo, J. R., Yang, L., & Wang, W. (2008). Estimating groundwater recharge in Hebei Plain, China under varying land use practices using tritium and bromide tracers. *Journal of Hydrology, 356*(1–2), 209-222. Retrieved from http://www.sciencedirect.com/science/article/pii/S0022169408001947. doi:http://dx.doi.org/10.1016/j.jhydrol.2008.04.011

Wang, J. (2003). Riverbank Filtration Case Study at Louisville, Kentucky. In C. Ray, G. Melin, & R. Linsky (Eds.), *Riverbank Filtration* (Vol. 43, pp. 117-145): Springer Netherlands.

Wang, Y. (2014). *Assessment of Ozonation and Biofiltration as a Membrane Pre-treatment at a Full-scale Drinking Water Treatment Plant*. UWSpace, Retrieved from http://hdl.handle.net/10012/8807

Wang, Y., Jiao, J. J., & Cherry, J. A. (2012). Occurrence and geochemical behavior of arsenic in a coastal aquifer–aquitard system of the Pearl River Delta, China. *Science of The Total Environment, 427-428*, 286-297. Retrieved from http://www.sciencedirect.com/science/article/pii/S0048969712004998. doi:https://doi.org/10.1016/j.scitotenv.2012.04.006

Weng, L., Temminghoff, E. J., Lofts, S., Tipping, E., & Van Riemsdijk, W. H. (2002). Complexation with dissolved organic matter and solubility control of heavy metals in a sandy soil. *Environmental science & technology, 36*(22), 4804-4810.

WHO. (2011). Guidelines for drinking-water quality. In (fourth ed., Vol. ISBN 978 92 4 154815 1, pp. 564). World Health Organization, Geneva, Switzerland.

Wintgens, T., Nättorp, A., Elango, L., & Asolekar, S. R. (2016). *Natural Water Treatment Systems for Safe and Sustainable Water Supply in the Indian Context: Saph Pani*: IWA Publishing.

Wünsch, U., Murphy, K., & Stedmon, C. (2017). The One-Sample PARAFAC Approach Reveals Molecular Size Distributions of Fluorescent Components in Dissolved Organic Matter. *Environmental Science and Technology, 51*(20), 11900–11908. doi:10.1021/acs.est.7b03260

Wünsch, U. J., Geuer, J. K., Lechtenfeld, O. J., Koch, B. P., Murphy, K. R., & Stedmon, C. A. (2018). Quantifying the impact of solid-phase extraction on chromophoric dissolved organic matter composition. *Marine Chemistry, 207*, 33-41. Retrieved from http://www.sciencedirect.com/science/article/pii/S0304420318301257. doi:https://doi.org/10.1016/j.marchem.2018.08.010

Yamashita, Y., Scinto, L. J., Maie, N., & Jaffé, R. (2010). Dissolved Organic Matter Characteristics Across a Subtropical Wetland's Landscape: Application of Optical Properties in the Assessment of Environmental Dynamics. *Ecosystems, 13*(7), 1006-1019. Retrieved from https://doi.org/10.1007/s10021-010-9370-1. doi:10.1007/s10021-010-9370-1

Yang, C., Li, S., Liu, R., Sun, P., & Liu, K. (2015). Effect of reductive dissolution of iron (hydr)oxides on arsenic behavior in a water–sediment system: First release, then adsorption. *Ecological Engineering, 83*, 176-183. doi:10.1016/j.ecoleng.2015.06.018

Yang, L., Chen, W., Zhuang, W.-E., Cheng, Q., Li, W., Wang, H., . . . Liu, M. (2019). Characterization and bioavailability of rainwater dissolved organic matter at the southeast coast of China using absorption spectroscopy and fluorescence EEM-PARAFAC. *Estuarine, Coastal and Shelf Science, 217*, 45-55. Retrieved from http://www.sciencedirect.com/science/article/pii/S0272771418307212. doi:https://doi.org/10.1016/j.ecss.2018.11.002

Yang, L., Guo, W., Hong, H., & Wang, G. (2013). Non-conservative behaviors of chromophoric dissolved organic matter in a turbid estuary: Roles of multiple biogeochemical processes. *Estuarine, Coastal and Shelf Science, 133*, 285-292. Retrieved from http://www.sciencedirect.com/science/article/pii/S027277141300406X. doi:https://doi.org/10.1016/j.ecss.2013.09.007

Yang, L., Shin, H.-S., & Hur, J. (2014). Estimating the Concentration and Biodegradability of Organic Matter in 22 Wastewater Treatment Plants Using Fluorescence Excitation Emission Matrices and Parallel Factor Analysis. *14*(1), 1771. Retrieved from http://www.mdpi.com/1424-8220/14/1/1771.

Zhang, B., Xian, Q., Lu, J., Gong, T., Li, A., & Feng, J. (2016). Evaluation of DBPs formation from SMPs exposed to chlorine, chloramine and ozone. *Journal of Water and Health, 15*(2), 185-195. Retrieved from http://dx.doi.org/10.2166/wh.2016.136. doi:10.2166/wh.2016.136

Zhao, C., Gao, S.-J., Zhou, L., Li, X., Chen, X., & Wang, C.-C. (2019). Dissolved organic matter in urban forestland soil and its interactions with typical heavy metals: a case of Daxing District, Beijing. *Environmental Science and Pollution Research, 26*(3), 2960-2973. Retrieved from https://doi.org/10.1007/s11356-018-3860-7. doi:10.1007/s11356-018-3860-7

Zhu, Y., Zhai, Y., Teng, Y., Wang, G., Du, Q., Wang, J., & Yang, G. (2020). Water supply safety of riverbank filtration wells under the impact of surface water-groundwater interaction: Evidence from long-term field pumping tests. *Science of The Total Environment,* *711,* 135141. Retrieved from http://www.sciencedirect.com/science/article/pii/S0048969719351332. doi:https://doi.org/10.1016/j.scitotenv.2019.135141

Zwolsman, J. J. G., & Bokhoven, A. J. v. (2008). Impact of summer droughts on water quality of the Rhine River—a preview of climate change? *Water Science Technolology, 56,* 44-55.

LIST OF ACRONYMS

ATP	Adenosine Triphosphate
ADL	Aswan Dam Lake
AHD	Aswan High Dam
BF	Bank Filtration
BIX	Biological Index
BP	Biopolymers
BDOC	Biodegradable Dissolved Organic Carbon
CDOC	Chromatographic Organic Carbon
DCW	Delft Canal Water
DCWW	Delft Canal water mixed with secondary treated WasteWater
DBPs	Disinfection By-Products
DOC	Dissolved Organic Carbon
DOM	Dissolved Organic Matter
D.O	Dissolved Oxygen
EfOM	Effluent Organic Matter
F-EEM	Fluorescence Excitation-Emission Matrix
PARAFAC–EEM	F-EEM coupled with Parallel Factor Analysis
FIX	Fluorescence Index
FA	Fulvic Acid
HMs	Heavy Metals
HPC	Heterotrophic Plate Counts

HMW	High Molecular Weight
HA	Humic Acid
HS	Humic Substances
HIX	Humification Index
HOC	Hydrophobic Organic Carbon
ICP-MS	Inductively Coupled Plasma, Mass Spectroscopy
ICS	Iron Coated Sand
LC-OCD/OND	Liquid chromatography with an on-line organic carbon detection
LOD	Limit of Detection
LMW	Low Molecular Weight
F_{max}	Maximum Fluorescence Intensity
NOM	Natural Organic Matter
N-DBPs	Nitrogenous DBP
NR	Nile River
NCTW	Non-chlorinated Tap Water
OM	Organic Matter
OMPs	Organic Micropollutants
OCD	Organic Carbon Detector
PARAFAC	Parallel Factor Analysis
PC	PARAFAC Component
PFFCA	Parallel Factor Framework-Clustering Analysis
FC	PFFCA Component

PAHs	Polyaromatic Hydrocarbons
RU	Raman Unit
RI	Redox Index
GRED	Renaissance Dam in Grand Ethiopia
WW	Secondary treated WasteWater effluent
$SUVA_{254}$	Specific Ultraviolet Absorbance
ST	Secondary Treated WasteWater
TOC	Total Organic Carbon
OP	Treated wastewater from oxidation ponds plant
TCC	Tucker's Congruence Coefficient
TY	Tyrosine
WEOM	Water Extractable Organic Matter
UV_{254}	UV-Absorbance at 254 nm
UVD	Ultraviolet Absorbance Detector

LIST OF TABLES

LIST OF FIGURES

ABOUT THE AUTHOR

Ahmed is a researcher at the water management department, TU Delft. He obtained his bachelor degree in chemistry from Aswan University (Egypt), postgraduate diploma in applied environmental geoscience from Assiut University (Egypt), a master of science in water resources management at ITC, University of Twente. He conducted his PhD research at TU Delft and IHE-Delft, focussed on contributing to the transfer of bank filtration to developing countries with arid environments.

Ahmed has participated in several projects in water resources management, including; groundwater modelling, remote sensing for resources management, and modelling of water quality. He also developed a remote-sensing based model (AquaSEBS) to estimate the evaporation rate over fresh and saline water bodies. He is now a member of a research group which investigates the transport of DNA-particles in natural water systems.

Awards and fellowships

- **MSc Scholarship** from the Qalaa Holdings Scholarship Foundation in Cairo, Egypt to study MSc in in Water Resources and Environment management at International Institute for Geo-Information Sciences and Earth Observation (ITC), University of Twente, Enschede (Netherlands) (Sep 2011-March 2013).

- **Short course Scholarship** from Netherlands Fellowship (NFP) to attend a short course (Applied Groundwater Modeling) at IHE-UNESCO (June 2014).

- **PhD Scholarship** from Netherlands Fellowship (NFP) to conduct PhD research at IHE-UNESCO (2015-2020).

- **ICCE Conference Award** from the Norwegian Chemical Society, Oslo, Norway to attend the conference (June 2017).

Journals publications

- Abdelrady, A., Sharma, S., Sefelnasr, A., Abogbal, A., & Kennedy, M. (2019). Investigating the impact of temperature and organic matter on the removal of selected organic micropollutants during bank filtration: A batch study. Journal of Environmental Chemical Engineering, 7(1), 102904.

- Abdelrady, A., Sharma, S., Sefelnasr, A., & Kennedy, M. (2018). The Fate of Dissolved Organic Matter (DOM) During Bank Filtration under Different Environmental Conditions: Batch and Column Studies. Water, 10(12), 1730.

- Abdelrady, A., Sharma, S., Sefelnasr, A., & Kennedy, M. (2020). Characterisation of the impact of dissolved organic matter on iron, manganese, and arsenic mobilisation during bank filtration. Journal of Environmental Management, 258, 110003.

- Abdelrady, A.; Timmermans, J.; Vekerdy, Z.; Salama, M.S. Surface Energy Balance of Fresh and Saline Waters: AquaSEBS. Remote Sens. 2016, 8, 583.

- Sayed A. Selim, Ali M. Hamdan, Ahmed Abdel Rady "Groundwater Rising as Environmental Problem, Causes and Solutions: Case Study from Aswan City, Upper Egypt", Open Journal of Geology, 2014.

- Ali Mohammed Hemdan, Ahmed Ragab Abdelrady, "Vulnerability of the groundwater in the quaternary aquifer at El Shalal Village area, Aswan, Egypt". Arabian Journal of geosciences, 2011.

- Abdelrady, A.; Sharma, S.; Sefelnasr, A.; El-Rawy, M.; Kennedy, M. Analysis of the Performance of Bank Filtration for Water Supply in Arid Climates: Case Study in Egypt. *Water* 2020, *12*, 1816.

- Abdelrady, A., Bachwenkizi, J., Sharma, S., Sefelnasr, A., & Kennedy, M. (2020). The fate of heavy metals during bank filtration: Effect of dissolved organic matter. Journal of Water Process Engineering, 38, 101563.

Journal articles submitted/in preparation

- A. Abdelrady and Ali Obeid. Hydrochemistry and Hydrogeology Aspects of Alluvial Aquifer in Aswan City, Egypt: In "Groundwater in the Nile River Valley", Abdelazim M. Negm and Mustafa El-Rawy (eds), Advances in Sciences, Technologies and Innovations Series, Springer, Dordrecht, The Netherlands, (under review).

- Abdelrady, A., Sharma, S., & Kennedy, M. (2020). Impact of organic matter on the behaviour of chemical pollutants during bank filtration: an overview. Environmental Pollution, (in preparation).

Conference proceedings

- Abdelrady, Sharma S, J. Bachwenkizi, Kennedy M. Impact of Dissolved Organic Matter on Heavy Metals Removal during River Bank Filtration: column studies. IWA natural organic matter NOM7 conference, Tokyo, Japan, 7-10 Oct. (2019).

- Abdelrady, Sharma S, Sefelnasr A, Abogbal A, Kennedy M. Role of redox conditions and organic matter in the removal of organic micropollutants during bank filtration. ICCE conference, Oslo, Norway, 18-22 June (2017).

- Abdelrady, Sharma S, Sefelnasr A, Abogbal A, Kennedy M. Characterization of the behaviour of dissolved organic matter during bank filtration using PARAFAC-EEM and LC-OCD techniques. ISMAR10 conference, Madrid, Spain, 20-23 May (2019).

- Abdelrady, Sharma S, Sefelnasr A, Abogbal A, Kennedy M. The role of organic matter in the release of iron and manganese during bank filtration. ISMAR10 conference, Madrid, Spain, 20-23 May (2019).

- Abdelrady, J. Timmermans and Z. Vekerdy, "Evaporation over fresh and saline water" in EGU conference, Vienna, Australia, 2013.

- Ali Mohammed Hemdan, Ahmed Ragab Abdelrady, "Vulnerability of the groundwater in the quaternary aquifer at El Shalal Village area, Aswan, Egypt", in 1st international conference and exhibition on sustainable water supply and sanitation, Cairo, Egypt 2010.

Netherlands Research School for the
Socio-Economic and Natural Sciences of the Environment

D I P L O M A

for specialised PhD training

The Netherlands research school for the
Socio-Economic and Natural Sciences of the Environment
(SENSE) declares that

Ahmed Abdelrady

born on 16 November 1985 in Aswan, Egypt

has successfully fulfilled all requirements of the
educational PhD programme of SENSE.

Delft, 10 November 2020

Chair of the SENSE board   The SENSE Director

Prof. dr. Martin Wassen   Prof. Philipp Pattberg

The SENSE Research School has been accredited by the Royal Netherlands Academy of Arts and Sciences (KNAW)

K O N I N K L I J K E   N E D E R L A N D S E
A K A D E M I E   V A N   W E T E N S C H A P P E N

The SENSE Research School declares that Ahmed Abdelrady has successfully fulfilled all requirements of the educational PhD programme of SENSE with a work load of 53.5 EC, including the following activities:

SENSE PhD Courses

o   Environmental research in context (2016)
o   Research in context activity: 'Creating Wikipedia information page in Arabic on 'Riverbank filtration treatment technique' and communicating accessible report on 'water resources management in Egypt' (2019)

Selection of Other PhD and Advanced MSc Courses

o   Multivariate Analysis, Wageningen University (2015)
o   Where is little data: How to estimate design variables in poorly gauged basins, IHE Delft (2015)
o   3D Geological and hydrological modelling with groundwater management aspects, Assiut University (2016)
o   Groundwater modelling, Assiut University (2017)
o   Groundwater in arid and semi-arid environments, IHE Delft (2017)
o   Popular Scientific Writing, TU Delft (2017)
o   Presenting scientific research, TU Delft (2017)
o   Managing the academic publication review process, TU Delft (2017)
o   Foundations of Teaching, Learning & Assessment, TU Delft (2017)
o   Scientific text processing with LaTex, TU Delft (2018)
o   Analytic Storytelling & Data Visualization - A practical approach, TU Delft (2018+2019)

Management and Didactic Skills Training

o   Supervising one BSc and three MSc students (2017-2020)
o   Organising PhD week, IHE Delft (2016)
o   Reviewing papers in impacted scientific journals (2019-2020)
o   Teaching in the MSc course "Unit Operations in Water Treatment (Coagulation, Sedimentation, Flotation and Filtration)" (2020)

Oral Presentations

o   *Impact of Dissolved Organic Matter on Heavy Metals Removal during River Bank Filtration: column studies.* IWA natural organic matter NOM7 conference,7-10 October 2019, Tokyo, Japan
o   *The role of organic matter in the release of iron and manganese during bank filtration.* 10th International Symposium on Managed Aquifer Recharge, 20-23 May, 2019, Madrid, Spain

SENSE coordinator PhD education

Dr. ir. Peter Vermeulen